The Geometry of Higher-Dimensional Polytopes

Gennadiy Vladimirovich Zhizhin
Russian Academy of Sciences, Russia

A volume in the Advances in Chemical and Materials Engineering (ACME) Book Series

IGI Global
DISSEMINATOR OF KNOWLEDGE

Published in the United States of America by
IGI Global
Engineering Science Reference (an imprint of IGI Global)
701 E. Chocolate Avenue
Hershey PA, USA 17033
Tel: 717-533-8845
Fax: 717-533-8661
E-mail: cust@igi-global.com
Web site: http://www.igi-global.com

Library of Congress Cataloging-in-Publication Data

Names: Zhizhin, G. V. (Gennadii Vladimirovich), author.
Title: The geometry of higher-dimensional polytopes / by Gennadiy Vladimirovich Zhizhin.
Description: Hershey, PA : Engineering Science Reference, [2019] | Includes bibliographical references.
Identifiers: LCCN 2018010794| ISBN 9781522569688 (hardcover) | ISBN 9781522569695 (ebook)
Subjects: LCSH: Polytopes. | Polygons. | Molecules--Models.
Classification: LCC QA691 .Z45 2019 | DDC 516/.158--dc23
LC record available at https://lccn.loc.gov/2018010794

This book is published in the IGI Global book series Advances in Chemical and Materials Engineering (ACME) (ISSN: 2327-5448; eISSN: 2327-5456)

British Cataloguing in Publication Data
A Cataloguing in Publication record for this book is available from the British Library.

Advances in Chemical and Materials Engineering (ACME) Book Series

ISSN:2327-5448
EISSN:2327-5456

MISSION

The cross disciplinary approach of chemical and materials engineering is rapidly growing as it applies to the study of educational, scientific and industrial research activities by solving complex chemical problems using computational techniques and statistical methods.

The **Advances in Chemical and Materials Engineering (ACME) Book Series** provides research on the recent advances throughout computational and statistical methods of analysis and modeling. This series brings together collaboration between chemists, engineers, statisticians, and computer scientists and offers a wealth of knowledge and useful tools to academics, practitioners, and professionals through high quality publications.

COVERAGE

- Biomaterials
- Chemical Engineering
- Thermo-Chemical Treatments
- Industrial chemistry
- Statistical methods
- Coatings and surface treatments
- Artificial Intelligence Methods
- Ceramics and glasses
- Wear of Materials
- Nanomaterials

IGI Global is currently accepting manuscripts for publication within this series. To submit a proposal for a volume in this series, please contact our Acquisition Editors at Acquisitions@igi-global.com or visit: http://www.igi-global.com/publish/.

The Advances in Chemical and Materials Engineering (ACME) Book Series (ISSN 2327-5448) is published by IGI Global, 701 E. Chocolate Avenue, Hershey, PA 17033-1240, USA, www.igi-global.com. This series is composed of titles available for purchase individually; each title is edited to be contextually exclusive from any other title within the series. For pricing and ordering information please visit http://www.igi-global.com/book-series/advances-chemical-materials-engineering/73687. Postmaster: Send all address changes to above address.

Titles in this Series

For a list of additional titles in this series, please visit:
https://www.igi-global.com/book-series/advances-chemical-materials-engineering/73687

Composites and Advanced Materials for Industrial Aplications
K. Kumar (Birla Institute of Technology, India) and J. Paulo Davim (University of Aveiro, Portugal)
Engineering Science Reference • ©2018 • 402pp • H/C (ISBN: 9781522552161) • US $225.00

Emerging Synthesis Techniques for Luminescent Materials
Ratnesh Tiwari (Bhilai Institute of Technology, India) Vikas Dubey (Bhilai Institute of Technology, India) and Sanjay J. Dhoble (Rashtrasant Tukadoji Maharaj Nagpur Universit, India)
Engineering Science Reference • ©2018 • 505pp • H/C (ISBN: 9781522551706) • US $215.00

Handbook of Research on Ergonomics and Product Design
Juan Luis Hernández Arellano (Autonomous University of Ciudad Juárez, Mexico) Aide Aracely Maldonado Macías (Autonomous University of Ciudad Juárez, Mexico) Juan Alberto Castillo Martínez (University of Rosario, Colombia) and Porfirio Peinado Coronado (Autonomous University of Ciudad Juárez, Mexico)
Engineering Science Reference • ©2018 • 446pp • H/C (ISBN: 9781522552345) • US $235.00

Mechanical Properties of Natural Fiber Reinforced Polymers Emerging Research and ...
Sarah S. Gebai (International University of Beirut, Lebanon) Ali M. Hallal (International University of Beirut, Lebanon) and Mohammad S. Hammoud (International University of Beirut, Lebanon)
Engineering Science Reference • ©2018 • 228pp • H/C (ISBN: 9781522548379) • US $165.00

Advanced Solid Catalysts for Renewable Energy Production
Sergio González-Cortés (Oxford University, UK) and Freddy Emilio Imbert (Univsersidad de Los Andes, Venezuela)
Engineering Science Reference • ©2018 • 520pp • H/C (ISBN: 9781522539032) • US $225.00

Production, Properties, and Applications of High Temperature Coatings
Amir Hossein Pakseresht (University of Tehran, Iran & Materials and Energy Research Center, Iran)
Engineering Science Reference • ©2018 • 557pp • H/C (ISBN: 9781522541943) • US $235.00

For an entire list of titles in this series, please visit:
https://www.igi-global.com/book-series/advances-chemical-materials-engineering/73687

IGI Global
DISSEMINATOR OF KNOWLEDGE

701 East Chocolate Avenue, Hershey, PA 17033, USA
Tel: 717-533-8845 x100 • Fax: 717-533-8661
E-Mail: cust@igi-global.com • www.igi-global.com

Table of Contents

Preface

In a recently published monograph G. Zhizhin, "Chemical Compound Structures and the Higher Dimension of Molecules: Emerging Research and Opportunities" (Zhizhin, 2017), the structures of the chemical compounds of s -, p -, d – and f - elements of the table of Mendeleev (Mendeleev, 1934) are systematically studied. It is shown that the molecules of most of them have a dimension greater than three. The simplest of these compounds are binary compounds formed by any two elements of the periodic system of elements. Even Fersman (Fersman, 1937) found that most of the binary compounds found in nature have structures from some limited set of structures. In the monograph (Zhizhin, 2017) it is proved that all these structures have a higher dimension (Table 1), i.e. their molecules form from a geometric point of view polytopes of higher dimension.

Table 1. Binary compounds of the elements

N	Type of the structure	The compounds d - and f elements with this type of the structure	The compounds s – and p - elements with this type of the structure
1	rock salt	MnO, FeO, CoO, NiO, CdO	LiF, NaF, KF, LiCl, NaCl, KCl, MgS, RbCl, LiBr, NaBr,KBr, RbBr, LiI, CaS, NaI, KI, MgO, CaO, SrO, BaO, SrS, BaS, PbS
2	Adamantane	Ag_2O, Cu_2O, ZnS, CuCl	Pb_4O_6, As_4O_6, Sb_4O_6, P_4O_4, $SiCl_4$, BeF_2, HgS
3	titanium chloride	TiCl	RbF, CsCl, CsF, CsBr, CsI
4	Rutile	MnF_2, CoF_2, NiF_2, CuF_2, ZnF_2, MnO_2, MoO_2	MgF_2, SnO_2, PbO_2
5	Wurtzite	ZnO, CdO, ZnS	BeO, AlN
6	Fluorite	CdF_2, $MnCl_2$, $FeCl_2$, $CoCl_2$, $NiCl_2$, $ZnCl_2$, $CdCl_2$, Cu_2S	CaF_2, SrF_2, BaF_2, RbF_2, AlF_8, $MgCl_2$, $CaCl_2$, $SrCl_2$, $BaCl_2$, $PbCl_2$, $AlCl_3$, $SiCl_4$, Li_2O, Na_2O, K_2O, Rb_2O, Cs_2O, Li_2S, Na_2S, K_2S, Rb_2S

The simplest compound of transition elements with three - dimensional structure has a cubic unit cell, for example, oxides of transition elements. Since such structures form chlorides, bromides and iodides of alkali metals, one will refer to these structures rock salt structures. Many binary chemical compounds have the structure as adamantane molecule. In work of Zhizhin (2014a) it was proved that the adamantane molecule consisting of 10 carbon atoms that make up the bulk of the unit cell of the diamond has of dimension 4. In compounds Ag_2O, Cu_2O at locations 10 of the carbon atoms are oxygen atoms, and atoms of copper and silver are arranged linearly between oxygen atoms. Many drugs are also a group of 10 carbon atoms as in the adamantane molecule. Among the inorganic and organometallic compounds have a number of structural analogues of adamantane (Table 1). All of these compounds have dimension 4 or even higher.

A series of binary compounds have a structure in the form of cube with centrum as in titanium chloride at which titanium ions are arranged in the centrum of the cube bat chlorine ions are arranged in vertices of the cube. Can to call these structures titanium chloride structure. Haw it is shown in work of Zhizhin & Diudea (2016) this structure has dimension 4.

A series of binary compound have a structure of the mineral rutile TiO_2. In this compound each titanium atom is surrounded by six the oxygen atoms in the octahedral coordination. To compounds with such structure to concern for example fluorides of copper, zinc, magnesium, manganese, cobalt, nickel. One will call these structures rutile structure. In work of Zhizhin & Diudea (2016) it is shown that octahedron with centrum have dimension 4. Therefore, all these structures have dimension 4. A series of binary compounds have structure of wurtzite - mineral ZnS, in which from compound ZnS with the structure of the adamantane zinc atom and sulfur atom have the tetrahedron coordination. The centrum each tetrahedron is vertex of another tetrahedron. The wurtzite structure have compounds ZnO, CdS, ZnS. The dimension of this structure defined in monograph Zhizhin (2017). This equal to 4. A series of binary compounds have a fluorite structure – mineral CaF_2 (fluorspar). Each calcium ion in this structure is in cube surrounded by fluorine ions, and each fluorine ion is in tetrahedron surrounded by calcium ions. The fluorite structure have for example chlorides of transition metals (Table 3). The dimension of this compounds also equal to 4 (Zhizhin, 2017).

Silicon, germanium, tin and lead also form a molecule of adamantane or a molecule that is topologically close to the adamantane molecule (Gillespie, 1972; Gillespie & Hargittai, 1991). The dimension of these molecules is 4 or more than four. Phosphorus, antimony, arsenic and bismuth form already considered polytopes for other elements: a tetrahedron with a center, an octahedron with a center, an adamantane molecule. All these molecules have a dimensionality of 4 or higher. Chlorine, bromine and iodine show the greatest possible numbers of oxidation states, interacting with previously considered elements. The dimensions of these compounds are often higher than three. Halogen compounds lead to molecules of higher dimensionality. For example, the iodine heptafluoride molecule IF_7 has the form of a 6 - simplex with a center whose dimension is 7. Noble gases in conjunction with other elements form molecules with tetrahedral coordination of higher dimension.

The complex molecules have even greater dimensions, for example, biomolecules (Zhizhin, 2016, 2017).

When the molecules interact with each other, their geometry plays an important role. Therefore, molecules are able to have higher dimensions, which influences how they interact with other molecules leading to advances in various fields including nanomedicine, nanotoxicology and quantum biology. Taking into account the important role of geometry in the interaction of higher - dimensional molecules, this monograph is devoted to the study of the geometry of polytopes of higher dimension that are models of molecules. It should be noted that previous studies of the geometry of polytopes (for example, Delone, 1961; Coxeter, 1963; Grunbaum, 1967; Zhigler, 1995; Zhizhin, 2014b) belonged to polytopes of dimension 3 and less often 4. At the same time, higher dimensions were not considered practically. In addition, these studies were of an abstract nature, i.e. they did not contact specific structures of chemical compounds.

The proof that many molecules have a form in the form of polytopes of higher dimension requires a detailed study of the geometry of these polytopes. This is the subject of this monograph. In addition, it should indicate the possibility of manifesting higher dimensions, not only in the structures of chemical compounds, but also in other areas of natural phenomena close to man. It is required to enumerate the geometric forms and features of these manifestations, i.e. to classify the forms of polytopes of higher dimension, not limited to polytopes of dimensionality 4. Based on the results of studies published in the recently published monograph of Zhizhin (Zhizhin, 2017), their further development and solution of a number of problems arising in these studies is necessary.

It is required to consider in detail the structures of the boundary complexes of polytopes of higher dimension, paying attention in these structures to the belonging of faces of lower dimension to faces of higher dimension. It is important, for example, to analyze the structure of chemical compounds formed by valence bonds.

It is required to consider groups of symmetry transformations of polytopes of higher dimension. Taking into account, that polytopes of smaller dimension enter in their boundary complexes, it is necessary that these studies be based on studies of the symmetry transformations of polytopes entering their boundary complexes.

Zhizhin (Zhizhin, 2017) found that polytopes, which are products of polytopes, are of great importance in the structures of chemical compounds. These polytopes are called polytopic prismahedrons. They do not belong to the class of simplicial polytopes, which are sufficiently well studied. Therefore, it is necessary to pay attention to the geometry of polytopic prismahedrons with allowance for the values of incidence of elements of different dimensions of their boundary complexes.

In particular, it is necessary to pay attention to a new type of polytopes, so - called poly-incident polytopes, in which simultaneously there are faces of some dimension incident to faces of higher dimensions, with different incidence values. Polytopes dual to polytopic prismahedrons are also of interest.

It is required to study in detail the structure of n – cross - polytopes, especially for large values of n. Since the announced, so - called Gosset polytope, used in various mathematical works, has a dimension of 8, and its boundary complex includes cross - polytopes of dimension 7. The question of the possibility of simultaneous presence in polytopes of high dimensionality of faces in the form of cross polytopes and polytopes of another form.

The proof of the existence of higher - dimensional clusters, cited in the works of Zhizhin and Dudea (Zhizhin & Diudea, 2016; Zhizhin, 2017; Diudea, 2018) requires the study of the hierarchical filling of a space of higher dimension and analysis of scaling processes in these fillings.

Until now, the eighteenth problem of Hilbert on the construction of spaces of higher dimension by means of congruent figures remains unsolved. In this context, this problem is posed as the construction of spaces of higher dimension with the help of polytopic prismahedrons.

To substantiate physically the possibility of the existence of molecules of higher dimension, the question of the existence of solutions of the Schrödinger equation, which describes the distribution of electrons in the vicinity of the atomic nucleus, in a space of higher dimension, is of considerable importance.

In the monograph Zhizhin (Zhizhin, 2017) it was first proved that the Schrödinger equation has a solution in four - dimensional space. It is required to solve the problem of the existence of solutions of the Schrödinger equation in spaces of higher dimension.

All these issues are reflected in this monograph, which consists of nine chapters.

In the Chapter 1 “Polytopes of higher dimension in the Nature”, three important examples of the manifestation of polytopes of higher dimension in Nature are considered. This chapter contains an analysis of the analysis of phase transitions of the second kind in condensed media. They are characterized by a change in the symmetry of the structure, for the explanation of experimental data, which requires the attraction of spaces of higher dimension. Changing the structure is accompanied by a scaling process, i.e. the change in the size of the unit cells of the structure during the phase transition. This leads in this way to the appearance in structures of polytopes of higher dimension. The Chapter 1 also contains an analysis of the structure of quasicrystals. It is shown that the diffraction patterns of quasicrystals can be considered as a projection onto a two - dimensional plane of the structure of a space of higher dimension. This reveals a hidden order in quasi - crystals, not previously detected. An elementary cell of the quasicrystal structure is constructed, which is a polytope of higher dimension. There is section devoted of the structure of the DNA molecule. For the first time, taking into account the higher dimensionality of the constituent DNA molecules, the geometrical cause of the DNA molecule twisting into a spiral is revealed.

In the Chapter 2 “Boundary complexes and interior points of the polytopes” structures of boundary complexes of regular n - dimensional polytopes: n - cubes, n - simplexes, n - cross-polytopes are considered. In addition, the structures of the boundary complexes of n -dimensional polytopes with center are considered. It is proved that regular n - dimensional polytopes have interior points, i.e. the boundary complexes do not occupy the entire volume of polytopes. If the n - dimensional polytopes have a vertex in the center, then the boundary complexes occupy the whole volume of the polytope. This is important for describing the clusters to which the last section of the chapter is devoted. The concept of a cluster surface of higher dimension corresponds to the notion of a polytope boundary complex. If the boundary complex occupies the entire volume of the polytope, this corresponds to the case when the cluster surface coincides with its volume. The important for further analysis property of boundary complexes of regular n - dimensional polytopes is proved. The incidence coefficient of elements of n – cross -polytopes is significantly higher than the corresponding incidence coefficients of elements of n -simplexes and n - cubes.

In the Chapter 3 "The number of symmetry transformation of convex regular polytopes in the n – space" symmetry transformations of regular polytopes of dimension n (n - cubes, n - simplexes of n - cross - polytopes) are considered, using the symmetry transformations of their facets. It is investigated how the symmetry transformations of the facets lead to symmetry transformations of the polytope under the condition of continuity of the polytope as a whole. It is proved that the number of transformations of a regular n - polytope is equal to the product of the number of symmetry transformations of the facets by the number of facets.

In the Chapter 4 "Polytopic prismahedrons" the structure of the product of the polytopes are considered. The products of the polytopes are polytopes that do not belong to the well-studied class of simplicial polytopes. Therefore, their research is of independent interest. In addition, the products of polytopes (polytopic prismahedrons) provide a solution to the 18th Hilbert problem in a space of higher dimension - constructing spaces of higher dimension from congruent figures. Analytical dependences of the characteristics of the product structure of polytopes as functions of the characteristics of the structures of the factors are obtained. The geometric properties of the boundary complexes of polytopic prismahedrons and their interior points are investigated. Polytopic prismahedrons can be the result of interaction of molecules of higher dimension. The possibility of simultaneous coordinated hierarchical and translational filling of n - dimensional spaces by polytopic prismahedrons is shown.

In the Chapter 5 "Poly – incident and dual polytopes" polytopes, dual to polytopic prismahedrons, are considered. It is shown that these polytopes form a new class of polytopes. In each polytope of this class elements of the same dimension simultaneously there are, but with different incidence values of elements of lower dimension with respect to elements of higher dimensionality. These polytopes of a new type are called poly - incident polytopes. The geometry of specific poly - incident polytopes and dual polytopes are studied, including for large dimensionality of polytopes. The existence of a previously unknown polytope consisting of 100 tetrahedrons is proved. Obtained his image with a list of all its components.

In the Chapter 6 "The detailed structure of n – cross – polytopes and polytopes with their participation "the structure of n – cross - polytopes at large n is investigated in detail. Confirmed by the construction that n – cross - polytopes can only be simplicial polytopes, although from the sets of tetrahedrons entering these polytopes can be distinguished by n – cross - polytopes of lower dimension. Polytopes, dual to n - cross-polytopes, are constructed, and new images of large dimension n - cubes are obtained. It is proved that n – cross - polytopes can be faces of polytopes of higher dimension if these polytopes are polytopic prismahedrons. A detailed analysis of the possibility of the existence of the Gosset (Gosset, 1900) polytope is carried out. According to Gosset's statements, this polytope should include 7 – cross - polytopes as facets. The number of facets of the simplex type and the cross - polytopes type specified by Gosset was adopted as the initial conditions. Taking into account the law on the change in incidence of faces, proved in this book, depending on their dimension and the type of polytope, it is established that the number of faces of various dimensions and types that Gosset points out for the polytope in his name contradict the law of incidence that was unknown to him. Thus, it is proved that the Gosset polytope does not exist.

In the Chapter 7 the hierarchical fillings of n - dimensional spaces with geometric figures are studied, accompanied by a change in their size (scaling). The scaling process in this case is determined only by the geometric characteristics of the figure. Two ways of hierarchical filling of space by figures are defined. The modes of hierarchical filling of spaces by plane, three -dimensional, four - dimensional and five - dimensional figures are investigated. The geometric progressions that characterize a discrete series of figures in the scaling process are calculated, depending on the shape of the figure and its dimension.

In the Chapter 8 it is proved that the stationary Schrödinger equation has a solution in principle for any dimension of the space around the nucleus of the atom. As an example, the solution of the stationary Schrödinger equation in five - dimensional space is given. It is shown that, as a result of the solution of the Schrödinger equation in p - dimensional space, the electron system in the vicinity of the nucleus is characterized by quantum numbers in the number $p + 1$: the principal quantum number, the orbital quantum number, the spin quantum number and the p - 2 magnetic quantum numbers.

In the Chapter 9 it is proved that polytopic prismahedrons have the necessary properties for partitioning the n - dimensional spaces of a face into a face. Consequently, they satisfy the conditions for solving the 18th Hilbert problem (Hilbert, 1901) of constructing n - dimensional spaces. An analytical method for constructing dimensional spaces from polytopic prismahedrons is developed. It is found that neighboring polytopic prismahedrons in these partitions can have common geometric elements of any dimension or do not have common elements. This refuted a priori the accepted condition in the theory of stereohedrons (Delone, 1961) about the necessity that adjacent stereohedrons of dimension n have common faces of dimension n - 1 or have no common elements.

It is shown that the set of lines arising in the hierarchical filling of spaces simultaneously allow translational symmetry. Therefore, simultaneous coordinated hierarchical and translational filling of spaces is possible.

A modern concept of describing the structure of matter is formulated. It is free of previously accepted limiting conditions and taking into account experimental data on the scaling process in various physical systems, as well as the formation of structures characteristic of quasi - crystals.

The subject of the book is the study of the geometry of polytopes of higher dimension, characteristic for molecules and chemical compounds established in the author's previous monograph "Chemical Compound Structures and the Higher Dimension of Molecules: Emerging Research and Opportunities" (Zhizhin, 2017). Since chemical compounds are the basis of the substance from which all living organisms, the environment, the bowels of the Earth and the whole noosphere are built, the subject of the book is important for various fields of knowledge: physics, chemistry, mathematics, biology, geology and various branches of these sciences.

The target audience of the book are students, university professors, researchers of research centers.

This book, together with the author's previous book "Chemical Compound Structures and the Higher Dimension of Molecules: Emerging Research and Opportunities" (Zhizhin, 2017), makes a significant contribution to the notion of the structure of matter and, in general, to the creation of a new scientific basis for modern natural science.

REFERENCES

Coxeter, H.S.M. (1963). *Regular Polytopes*. London: John Wiley & Sons, Inc.

Delone, B. (1961). Proof of the main theorem of the theory of stereohedrons. *Reports of the Academy of Sciences of the USSR*, *138*(6), 1270–1272.

Diudea, M. V. (2018). Hypercube Related Polytopes. *Iranian J. Math. Chem.*, *9*(1), 1–8.

Fersman, A. E. (1937). *Geochemistry* (Vol. 3). Leningrad: ONTI.

Gillespie, R. J. (1972). *Molecular Geometry*. New York: Van Nostrand Reinhold Company.

Gillespie, R. J., & Hargittai, I. (1991). *The VSEPR Model of Molecular Geometry*. London: Allyn & Bacon.

Gosset, T. (1900). On the regular and semi-regular figures in space of n-dimensions. *The Messenger of Mathematics*, *29*, 43–48.

Grunbaum, B. (1967). *Convex Polytopes*. London: Springer.

Mendeleev, D. I. (1934). *The periodic law of chemical elements*. Moscow, Leningrad: Gos. Chemical and Technical Publishing.

Zhizhin, G. V. (2014a). On the higher dimension in nature. *Biosphere*, *6*(4), 313–318.

Zhizhin, G. V. (2014b). *World – 4D*. St. Petersburg: Polytechnic Service.

Zhizhin, G. V. (2016). The structure, topological and functional dimension of biomolecules. *J. Chemoinformatics and Chemical Engineering*, *5*(2), 44–58.

Zhizhin, G. V. (2018). *Chemical Compound Structures and Higher Dimension of Molecules: Emerging Research and Opportunities*. Hershey, PA: IGI Global. doi:10.4018/978-1-5225-4108-0

Zhizhin, G. V., & Diudea, M. V. (2016). Space of Nanoworld. In M. V. Putz & M. C. Mirica (Eds.), *Sustainable Nanosystems, Development, Properties, and Applications* (pp. 214–236). Hershey, PA: IGI Global.

Zhizhin, G. V., Khalaj, Z., & Diudea, M. V. (2016). Geometrical and topology dimensions of the diamond. In A. R. Ashrafi & M. V. Diudea (Eds.), *Distance, symmetry and topology in carbon nanomaterials* (pp. 167–188). London: Springer. doi:10.1007/978-3-319-31584-3_12

Ziegler, G. M. (1995). *Lectures on Polytopes*. Springer. doi:10.1007/978-1-4613-8431-1

Chapter 1
Polytopes of Higher Dimension in the Nature

ABSTRACT

Areas of research into the phenomena of nature in which the influence of polytopes of higher dimension is described in this chapter. These include studies of the structures of many chemical compounds whose molecules exhibit the properties of polytopes of higher dimension. This leads to the creation of higher-dimensional stereochemistry. Phase transitions of the second kind are accompanied by a change in the symmetry of the structure of matter, the description of which, in agreement with the experimental data, requires the attraction of spaces of higher dimension. Elementary cells of quasicrystals, having the form of polytopic prismahedrons, are given (polytopes of higher dimension). The structure of DNA as sequence of the higher dimensional polytopes are given.

INTRODUCTION

Visual image of any object on the retina of the human eye is two - dimensional. For this reason, any object is perceived initially as a two - dimensional one. Similarly, we touch only the surface of objects, that is, the palpable image of objects has again two dimensions. The idea of three -dimensionality of an object forms only as a result of the comparison of mismatched images in the right and left eye, this difference being observed in motion. Comparison is a result of thinking and judgment appears as a result of it.

DOI: 10.4018/978-1-5225-6968-8.ch001

Therefore, even the three - dimensional conception of objects is an abstract idea and, therefore, it is difficult for many people. The way of abstraction leads us to the concept of space, i.e. to a conceptual space (space studied by geometry - geometrical space, (Poincare, 1985)). This conceptual space arose on the basis the perception of space, i.e. it based on the perceptual space (the space of ideas (Poincare, 1985)). A perceptual space itself is a reflection of the real material space. Perceptual space is an image of the real space, which exists regardless of whether we perceive it or not, and on what the perception we get from it (Coleman, 1970). However, given that perceptual space is the result of mental activity, the image of the real space is ambiguous. It may take into account the different human views about surrounding world, as well as either properties of the world that seem the most important to a person in the given concrete circumstances. It is also the reason for co - existence of different geometries. It should be borne in mind that the geometrical axioms are neither a priori synthetic judgments nor experimental facts. They are conventional provisions (agreement): choosing between all possible agreements, we are guided by experimental facts, but the choice is free and limited only by the need to avoid any contradiction "(Poincare, 1985).

Historically two main concepts of the space formed within which there were a lot of modifications of these concepts and related geometries. According to the first of these concepts associated with the names of Aristotle and Leibniz, the real space is a property of position of material objects. It is inextricably linked with the matter. The development of these ideas led to the well- known philosophical theses that there is no space without matter, just as there is no matter without space, and that space is a form of existence of the matter.

According to the second of these concepts, associated with the names of Democritus and Newton, space is the receptacle of all the material objects having no influence on space (Einstain, 1930). Exactly this idea of space was determinative for many centuries and serve as philosophical foundation of Euclidean geometry (Euclid, 2012). According to this view geometric space is continuous, endless, three - dimensional, uniform (all points of space are identical to each other), isotropic (all lines passing through a point are identical to each other).

Ideas of Leibniz in his time did not have any support, because they lead to ambiguity of Euclidean geometry. N.I. Lobachevsky's discovery of non-Euclidean noncontradictory geometry according to which through the point outside straight line more than one parallel lines can be drawn on the plane initiated rapid development of geometry. It is important that N. I. Lobachevsky associated geometry with physics and space with matter

(Lobachevsky, 1945). The works of Lobachevsky at non - Euclidean geometry it were continued by B. Riemann. In his famous lecture "On the hypotheses that lie at the foundations of geometry" (Riemann, 1854) he introduces the concept of n - extended manifold. It is essential that the n – dimensional extension it is determined by B. Riemann without introduction of infinite spaces. Moreover, the infinity of space is obviously contrary to the views of B. Riemann on n - dimensional extension. Definition it B. Riemann initially considers a finite region, and as a sign of manifold n - dimension manifold the position in this manifold is characterized by a change in n - dimensional ("simple") of extended values. B. Riemann showed that a manifold of (n + 1) - dimensions is a manifold of 1 - st dimension, consisting of elements (pixels) n - dimensions. Three - dimensional space can be considered as a special case of n - dimensions manifold.

On the basis of the above representations, the geometric shapes of many molecules were investigated (Zhizhin, 2017). The study showed that almost all elements of the periodic system of Mendeleev to announce in the chemical compounds the valence is higher than the valence determined by the number of the group in the periodic system. To explain this in the scientific literature resorts to assumptions about various mechanisms of electron interaction. Practical all these compounds form molecules of higher dimension (polytopes of higher dimension), determined by the Euler - Poincaré equation (Poincare, 1895). The existence of closed objects of higher dimension in a space of lower dimension does not contradict Riemann's geometry, which assumes the boundedness of a space with a given dimension. Images of many molecules of higher dimension are given. In particular, new images of biomolecules having high dimensional values were obtained.

In this chapter, as an illustration of the importance of polytopes of higher dimension in describing nature, three complex objects are selected: structures that arise in condensed media during phase transitions of the second kind; structure of quasicrystals and DNA molecules.

GEOMETRIC FEATURES OF PHASE TRANSITIONS OF THE SECOND KIND

The behavior of systems in the vicinity of the critical point - the point of phase transition of the second order has been studied by researchers for several decades (Fisher, 1965; Patashinskii & Pokrovsky, 1975; Izyumov &

Syromyatnikov, 1984). Different models of the media in the vicinity of the critical state were built: classic continual gas, Heisenberg model of magnetism, binary alloys, lattice gas, Ising`s model of magnetism (Fihser, 1965). When building the model in each case the Hamiltonian of the system characterizing its energy was recorded. It was found, in particular, that the models of ferromagnetic and lattice gas in the vicinity critical point are equivalent. To calculate the changes in the characteristics of the system (heat capacity, spontaneous magnetization, etc.) the series resolution of the Hamiltonian in the vicinity of the critical point according to the powers of temperature T deviation from the critical temperature T_C was used. Theoretically and experimental the singular behavior of environment characteristics at the phase transition of the second kind was discovered. Landau was the first (Landau, 1937) who linked the second order phase transition with breaking symmetry of the body structure at the point of phase transition, i.e. with a decrease of the order of the group symmetry structure. He introduced into consideration the order parameter characterizing the phase transition. Subsequently, it was found that the order parameter can have several components, i.e. it is a vector (Landau & Lifshitz, 1964). The dimension of this vector coincides with the dimension of irreducible representations of the symmetry group of high - symmetry phase. The idea of Kadanov (Kadanov, 1966) about the merger of unit cells (polytopes) in the high - symmetry phase into larger blocks during the phase transition, acting in a new phase as the new elementary cells was very productive in the study of critical phenomena. This idea (the idea of scaling) was confirmed later experimentally and it allowed Wilson (Wilson, 1971 a, b) and Fischer (Fisher, 1972) to introduce for consideration to describe the continuous phase transition system of differential equations, the so - called renormalization group equations. The fact of changes in the size of the unit cell (polytope) in the process of phase transition became the basis of the principle of similarity in phase transitions, expressed in a power law of the thermodynamic quantities change in phase transitions process (Fisher, 1965; Patashinskii & Pokrovsky, 1975; Izyumov & Syromyatnikov, 1984), the exponents of which are called critical indices. It was found that the equations of renormalization group essentially depend on dimension of the space in which the Hamiltonian of the system is studied. At that the space dimension i equal to 4, as it turns out, has a special significance, as critical indices, following from the theory of Landau coincide with the indices following from the equation of the renormalization group at $i = 4$, and don`t coincide when the space dimension equals to 3. Since, according to Landau's theory, the dimension of the phase is 4, we are forced to consider a polytope of dimension 4 as the unit cell, i.e. polytope of higher dimension. In this regard, Wilson proposed to investigate the renormalization - group equations near the selected value of $i = 4$, assuming the parameter $\varepsilon = i - 4$ to be small. Then you can resolve the Hamiltonian in series according to powers ε, and then return in the received results back to the familiar three dimensions, setting $\varepsilon = 1$. At that it is important at what fixed points the phase transition process comes to end. However, the peculiarities of achievement of fixed points of equations system of phase transitions renormalization group especially for large dimension of the order parameter have been investigated not enough, both theoretically and the more-experimental.

In this section, the main attention is paid in particular to the qualitative study of systems of differential equations of the renormalization group of phase transitions with order parameter of higher dimensions. The use of modern methods of qualitative research systems of differential equations in multidimensional phase spaces allowed receive principally new results about the nature of phase transitions drive for fixed points.

The System of Differential Equations of Phase Transitions

Renormalization Group

The energy of high - symmetry phase system in the point of phase transition vicinity, according to the theory of Landau, is written in the form of the Hamiltonian resolution in row to powers of order parameter. In this row the terms with spatial derivatives taking into account, in the general case, uneven distribution of the order parameter of the space shall be added, as well as terms associated with complexity of order parameter. This energy equation is called the Hamiltonian of Ginzburg - Landau. Each term in the row has a coefficient u_i, which is a function of temperature and, in the general case, of pressure. The functions u_i will ultimately represent the space phase coordinates of a system of phase transitions renormalization group differential equations. According to the ideas Kadanov – Wilson it`s necessary to carry out the transition from the initial Hamiltonian H to subsequent Hamiltonians H ‘, H’ ‘, H’ ‘’ ... (and, correspondingly, to the recursive variable $u_i', u_i'', u_i''' ...$) to describe the process of integration of units. The fixed point of this process will meet the necessary condition of the invariance of the Hamiltonian that corresponds, considering the operator of the transition to be infinitesimal, to the equilibrium position of the of differential equations system. In the vicinity of the fixed point, according to the method of Wilson - Fisher the resolution of the Hamiltonian in row according to powers of the parameter ε limited to a small number of members due to of assumptions about its smallness is used. The result is a system of differential equations for the recursive variables, in which superscripts already can be omitted. The number of variables in the resulting system is equal to or less than the number of irreducible representations of the symmetry group of high-symmetry phase, i.e. the dimension of the order parameter is equal to or greater than the number of variables in a system of differential equations of phase transitions renormalization group (Mukamel, 1975; Mukamel & Krinsky, 1976). The systems with dimensions of order parameter more than 3 are of particular interest, as these cases correspond to phases symmetry space group (Mukamel, 1975). The system of differential equations of the renormalization group in this case is more than three equations and for their the methods of qualitative research equations shall be applied in multidimensional phase spaces which were developed for solving problems in the mechanics of liquid, gas and plasma (Anderson, 1963; Zhizhin, 1972; Bogoyavlensci, 1980; Zhizhin, 2004; Zhizhin, 2005).

In general, the system of differential equations of the renormalization group of phase transitions has the form

$$\frac{du_j}{dt} = f_j\left(u_1, u_2, \ldots, u_n\right), j = 1 \div n. \tag{1}$$

This system of equations is autonomous because of the independent variable *t* is not included in the right parts of this equations. An independent variable can be represented by a dimensionless characteristic block size or dimensionless (normalized) time. From the general form of the system (1) one cannot a priori draw a conclusion about its features, in particular, about the types of equilibrium positions, about the kinds of aggregates of equilibrium positions, about the possibilities of transition from one position of equilibrium to another equilibrium position. It is necessary for this aim to consider the system (1) with specific functions f_j types.

Research of Phase Transitions Renormalization Group Equations With Four-Dimensional Order Parameter

A large number of substances in which phase transitions occur with four-dimensional order parameter are indicated (Mukamel, 1975; Shapiro & Axe & Shirane & Raccah, 1974; Mukamel & Krinsky, 1976). These include the structural phase transitions in NbO_2, magnetic phase transitions in TbAs, TbP, DyC_2 and others. A detailed derivation of phase transitions renormalization group equations with four-dimensional the order parameter is given (Mukamel, 1975). For example, a phase transition occurs in NbO_2 at T_C ~ 800^0 C. At that the high-symmetry phase having symmetry group $P4_2$ / mnm, lowers the symmetry group to $I4_1$ / a.

The Hamiltonian of the system with four - dimensional order parameter has the form

$$H = \int \left\{ \frac{1}{2} u_4 \sum_{j=1}^{4} \varphi_j^2 + \frac{1}{2} \sum_{j=1}^{4} \left(grad \varphi_j \right)^2 + u_1 \left(\sum_{j=1}^{4} \varphi_j^2 \right)^2 + u_2 \sum_{j=1}^{4} \varphi_j^4 + u_3 \left(\varphi_1^2 \varphi_2^2 + \varphi_3^2 \varphi_4^2 \right) \right\} dV \tag{2}$$

φ_j are components of the order parameter.

Using the technique of Wilson`s ε - resolution a system of equations for the recursive variables $u_j^{/}$ was received (Mukamel, 1975) of which three equations can be written restricting oneself to the terms of first order in ε, in the form of a closed autonomous system of differential equations

$$\frac{dv_1}{dt} = v_1 - 48v_1^2 - 24v_1v_2 - 4v_1v_3, \tag{3}$$

$$\frac{dv_2}{dt} = v_2 - 36v_2^2 - 48v_1v_2 - v_3^2, \tag{4}$$

$$\frac{dv_3}{dt} = v_3 - 48v_1v_3 - 24v_1v_3 - 8v_3^2, \tag{5}$$

where v_j is dimensionless scaled function u_j (j = 1, 2, 3), t is independent dimensionless variable.

In accordance with the system (3) - (5) in the phase space (v_1, v_2, v_3) there are five zero surfaces at points which derivatives $\frac{dv_j}{dt} = 0$:

The surface $F_1 : v_1 = 0$, in its points $\frac{dv_1}{dt} = 0$;

The surface $F_2 : v_1 = \frac{1}{48} - \frac{1}{2}v_2 - \frac{1}{12}v_3$, in its points $\frac{dv_1}{dt} = 0$;

The surface $F_3 : v_1 = \frac{v_2 - 36v_2^2 - v_3^2}{48v_2}$, in its points $\frac{dv_2}{dt} = 0$;

The surface $F_4 : v_3 = 0$, in its points $\frac{dv_3}{dt} = 0$;

The surface $F_5 : v_1 = \frac{1}{48} - \frac{1}{2}v_2 - \frac{1}{6}v_3$, in its points $\frac{dv_3}{dt} = 0$.

The points of intersection of zero surfaces in which simultaneously $\frac{dv_1}{dt} = \frac{dv_2}{dt} = \frac{dv_3}{dt} = 0$ are the positions of equilibrium of the system (3) – (5), i.e. its immovable points. Two zero surfaces F_1 and F_2 coincide with coordinated flats $v_1 = 0, v_3 = 0$. On the axis v_2 which is the intersection of zero surfaces F_1 and F_4 two positions of equilibrium are located. Their coordinates can be easily found if in equation of surface F_3 we assume that $v_1 = 0$ and $v_3 = 0$, i.e. there are immovable points:

1. $v_1 = 0, v_2 = 0, v_3 = 0$ and
2. $v_1 = 0, v_2 = \frac{1}{36}, v_3 = 0$.

The positions of equilibrium on open part of the coordinate plane $v_1 = 0$ are found as general points of intersection of the surfaces F_3, F_5 with this plane. Thus, from the system

$$v_3^2 = v_2\left(1 - 36v_2\right), v_2 = \frac{1}{24} - \frac{1}{3}v_3, \tag{6}$$

can to find coordinates of two more immovable points

3. $v_1 = 0, v_2 = \dfrac{1}{72}, v_3 = \dfrac{1}{12}$, and
4. $v_1 = 0, v_2 = \dfrac{1}{40}, v_3 = \dfrac{1}{20}$.

On the coordinate plane $v_3 = 0$ there are the already found points 1), 2) and besides the immovable point 5) with coordinates $v_1 = \dfrac{1}{48}, v_2 = 0, v_3 = 0$. The system does not have another the immovable points. It is essential that they are not present in the open part of the phase space, since the equations of surfaces F_2 and F_5 inconsistent in this area.

Linearizing the system of equations (3) - (4) in a small deviation from the equilibrium positions of the corresponding discriminants we find the eigenvalues of the equilibrium positions at the point:

1. $\lambda_1 = \lambda_2 = \lambda_3 = 1$; at the point
2. $\lambda_1 = \lambda_2 = \dfrac{1}{3}, \lambda_3 = -1$; at the point
3. $\lambda_1 = \dfrac{1}{3}, \lambda_2 = \dfrac{\sqrt{7} - 2}{3}, \lambda_3 = -\dfrac{\sqrt{7} + 2}{3}$; at the point
4. $\lambda_1 = \dfrac{1}{5}, \lambda_2 = \dfrac{\sqrt{41} - 6}{5}, \lambda_3 = -\dfrac{\sqrt{41} + 6}{5}$.

Thus, the equilibrium position 1) - 4) in three - dimensional phase space are unstable, since each of them has positive eigenvalues. The only point at which the linear approximation there are no positive eigenvalues of this point is 5). In this point are $\lambda_1 = -1, \lambda_2 = 0, \lambda_3 = 0$. It is clear that this point is doubly degenerated. Since the system (3) - (5) in the vicinity of this point in the linear approximation reduces itself to one equation $\dfrac{dv_1}{dt} = -\left(v_1 - \dfrac{1}{48}\right)$, the value $\lambda_1 = -1$ characterizes the movement to the point 5) only on the axis v_1. Therefore, it is necessary to consider the behavior of trajectories in three-dimensional vicinity of point 5), without limiting to the linear approximation and to consider the behavior of the trajectories in the phase space (v_1, v_2, v_3) as a whole. The statement that taking into account in the resolution of the Hamiltonian of terms of the smallness second order in ε withdraws the degeneracy of the singular points (see, for example, (Bogoyavlensci, 1980)), is incorrect. This only takes the system from "rough" state to "not rough" one and any small "stirring" of the system returns to its "rough" condition. In order to understand the behavior of the system in the vicinity of a singular point degenerate in the linear approximation one should simply give up the linear approximation for the deviation from the singular point.

To construct a three - dimensional phase portrait of a system (3) - (5) can to construct zero surfaces in the space (v_1, v_2, v_3) and define the line of intersection with the formation of fixed points of the system (Figure 1).

Under each of zero surfaces F_1, F_2, F_3 the corresponding phase variable increases, as indicated by the arrows at the zero surfaces. Surfaces F_3 and F_5 are crossed by lines connecting the points 3), 4) with the point 5). Above of zero surface corresponds phase coordinates only decrease. It is essential that the zero surface F_2 has only one point common with zero surface F_3 -

Figure 1. The zero surfaces of the system (3) – (5)

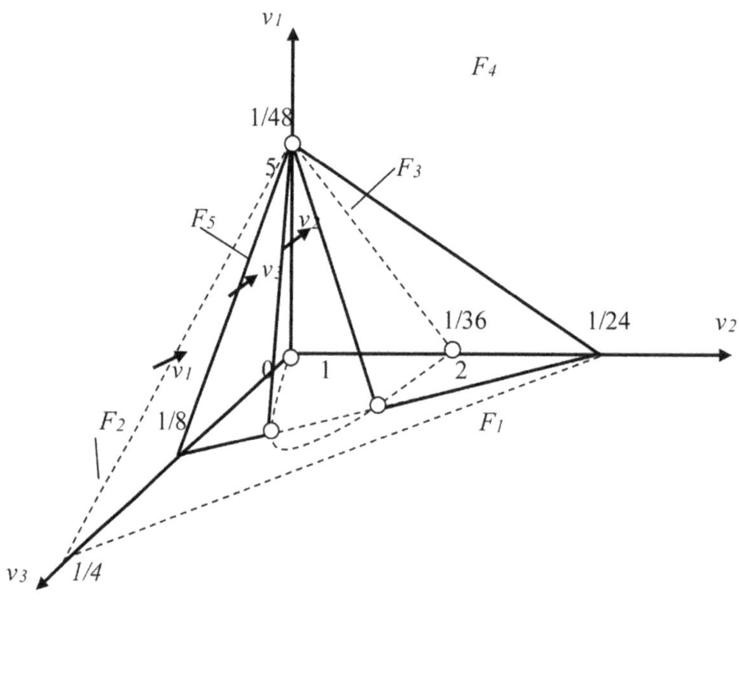

point 5) and intersects with the zero surface F_5 on straight line $v_1 = \frac{1}{48} - \frac{1}{24} v_2$, located on the zero surface F_4 and passing through the point 5). Based on the distribution of signs derivatives from the phase coordinates of the areas formed by the intersection of zero surfaces, and the analysis of equilibrium positions of the system the phase diagram of the system (3) - (5) shown in Figure 2 is derived.

It should be noted that, as follows from (3) - (5) coordinate planes $v_1 = 0, v_3 = 0$ are integral surface, i.e. they are phase portrait of the system, respectively at $v_1 = 0$ and $v_3 = 0$. Coordinate plane $v_2 = 0$ is not an integral one and trajectory, crossing her come out of the positive octant into the negative (for v_2) octant of the phase space.

Integral surface $v_3 = 0$ (see Figure 3) has an unstable point 2) of the saddle type, stable point 5) of node type and unstable point 1) the of node type.

The separatrix coming out from the saddle 2) goes into the node 5) also as a separatrix. Thus, all the solutions in the plane $v_3 = 0$ coming out from the fixed point 1) of node type belong to a fixed point 5). Integral surface $v_1 = 0$ (see Figure 4) has the unstable fixed point 1) the node type, the unstable

Figure 2. Trajectories of the system (3) - (5)

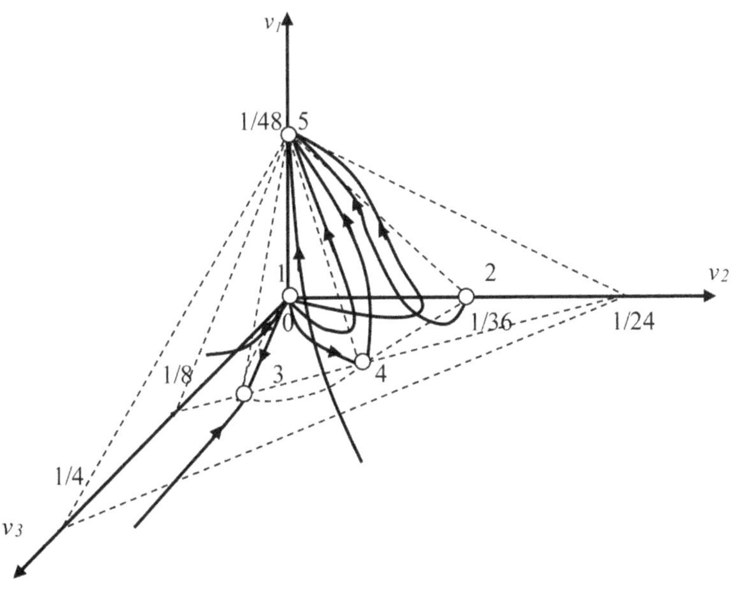

Figure 3. Trajectories of coordinate flat (v_1, v_2)

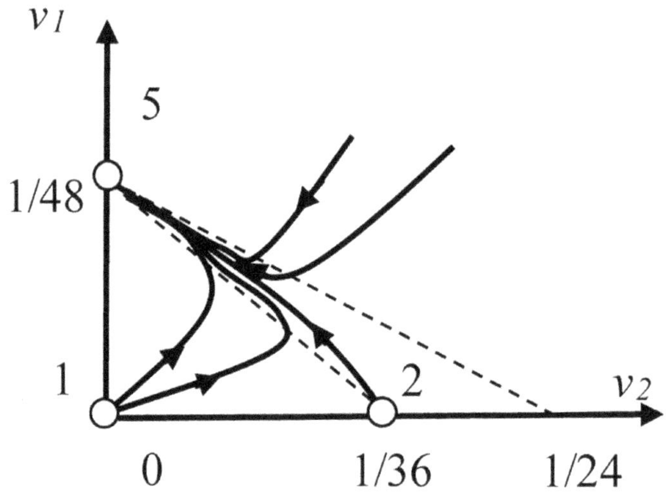

Figure 4. Trajectories of coordinate flat (v_3, v_2)

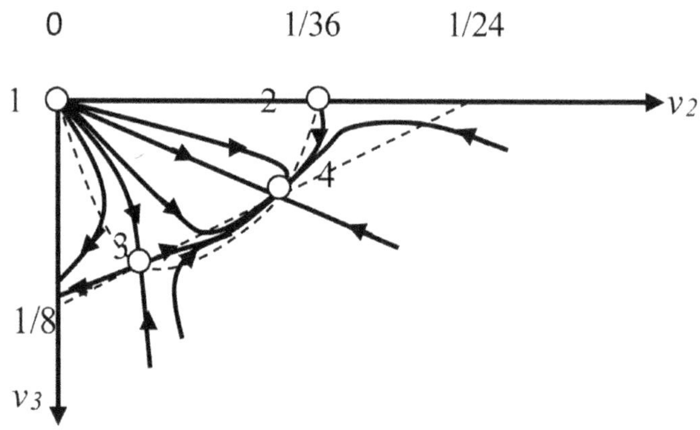

fixed point 2) of saddle type, the unstable fixed point 3) of saddle type and the stable fixed point 4) of node type.

In the open part of the phase space the trajectory located in the vicinity of integral surfaces qualitatively repeated the patterns of trajectories movement on these surfaces. It is essential that the fixed point 4) stable on the surface of the integrated $v_1 = 0$, becomes unstable (saddle - node) in the space (v_1, v_2, v_3), since it comes out of the separatrix entering the point 5). Last point remains stable in space - three-dimensional stable node. Solutions exiting unstable three-dimensional node 1), suitable to point 4) or 2) (saddle - node) and then turning, go to point 5). Solutions that go from point 1), most of them go into the negative octant of the phase space.

Thus, a qualitative study of solutions of (3) - (5) in three - dimensional phase space (v_1, v_2, v_3) proves that the fixed point 5) is stable not only in the linearization solutions in its vicinity. Moreover, and this is essential, point 5) is stable and on integral surface $v_1 = 0$ and on the whole three - dimensional phase space (v_1, v_2, v_3), i.e. solutions on different assumptions regarding the variables v_1, v_2, v_3 end to it. It's principal thing, as in (Mukamel, 1975) with reference to the (Aharony, 1973) it was claimed that a stable fixed point changes its type and location with the change of formulation of the problem from $v_3 = 0$ to the problem with $v_3 \neq 0$. The study of fixed points in (Mukamel, 1975; Aharony, 1973) was carried out with the accuracy up to terms of the smallness ε^2. This led to the "replication" of point 5): instead of one point 5) in the "rough" system appeared 4 points 5) "structurally unstable" system, as a consequence of incorrect statement of the problem. In addition, as can be detected by substitution, the coordinates of fixed points in (Mukamel, 1975), under the condition of terms of the second order smallness in ε do not satisfy the condition of equilibrium of the system considered in (Mukamel, 1975).

Investigation of the Equations of Phase Transitions Renormalization Group With Six-Dimensional Order Parameter

Phase transitions with a six-dimensional the order parameter can occur in a number of substances (Patashinskii & Pokrovsky, 1975; Mukamel & Krinsky, 1976a, b). They include, for example, phase transitions in $TbAu_2$, DyC_2, $NdTbO_2$, K_2IrCl_6 (space group *Fm3m*) magnetic phase transition occurs with the formation of the antiferromagnetic structure of type 111 (Mukamel & Krinsky, 1976a) occur. The Hamiltonian of the Ginzburg - Landau with six-dimensional order parameter has the form (Mukamel & Krinsky, 1976a, b)

$$H = \int \left\{ \frac{1}{2} u_4 \sum_{j=1}^{6} \varphi_j^2 + \frac{1}{2} \sum_{j=1}^{6} \left(grad\varphi_j \right)^2 - u_1 \sum_{j=1}^{6} \varphi_j^4 + u_3 \begin{pmatrix} \varphi_1^2\varphi_2^2 + \varphi_1^2\varphi_5^2 + \varphi_1^2\varphi_3^2 + \varphi_4^2\varphi_6^2 + \varphi_1^2\varphi_6^2 \\ +\varphi_4^2\varphi_3^2 + \varphi_3^2\varphi_2^2 + \varphi_5^2\varphi_6^2 + \varphi_2^2\varphi_6^2 + \varphi_3^2\varphi_5^2 \end{pmatrix} \right\} dV. \tag{7}$$

Using the technique of ε -resolution of Wilson (Mukamel & Krinsky, 1976b), a system of equations for the recursive variables $u_j^{/}$ was received. Three equations of this system can be written, restricting oneself to the terms of first order in ε, as a closed autonomous system of differential equations (Mukamel & Krinsky, 1976b). [In this case the equilibrium position of the system were also considered in the linear approximation in ε]

$$\frac{dv_1}{dt} = v_1 - 36v_1^2 - v_2^2 - 4v_3^2, \tag{8}$$

$$\frac{dv_2}{dt} = v_2 - 8v_2^2 - 24v_1v_2 - 8v_3^2, \tag{9}$$

$$\frac{dv_3}{dt} = v_3 - 24v_1v_3 - 4v_2v_3 - 12v_3^2. \tag{10}$$

In accordance with the system (8) - (10) in the phase space (v_1, v_2, v_3) there are four zero surface at the points where the derivatives are zero:

The surface $F_1 : v_1^2 - \frac{1}{36}v_1 + \frac{1}{36}v_2^2 + \frac{1}{9}v_3^2 = 0$, in its points

$$\frac{dv_1}{dt} = 0; \tag{11}$$

The surface $F_2 : v_1 = \frac{v_2 - 8v_2^2 - 8v_3^2}{24v_2}$, in its points

$$\frac{dv_2}{dt} = 0; \tag{12}$$

The surface $F_3 : v_1 = \frac{1}{24} - \frac{1}{2}v_3 - \frac{1}{6}v_2$, in its points

$$\frac{dv_3}{dt} = 0; \tag{13}$$

The surface $F_4 : v_3 = 0$, in its points $\frac{dv_3}{dt} = 0$.

The points of intersection of zero surfaces in which both $\frac{dv_1}{dt} = \frac{dv_2}{dt} = \frac{dv_3}{dt} = 0$ are positions of equilibrium of the system (11) - (13), i.e. its fixed points. The zero surface F_4 coincides with the coordinate plane $v_3 = 0$. The equilibrium positions are on it:

1. $v_1 = 0, v_2 = 0, v_3 = 0;$
2. $v_1 = \frac{1}{40}, v_2 = \frac{1}{20}, v_3 = 0;$
3. $v_1 = \frac{1}{36}, v_2 = 0, v_3 = 0;$
4. $v_1 = \frac{1}{72}, v_2 = \frac{1}{12}, v_3 = 0$

Furthermore, unlike the system with the four-dimensional order parameter, in this case in the open part of the phase space there are four more equilibrium positions, which belong to the zero surfaces F_1, F_2, F_3. Let`s describe in short the sequence in which they are located. From (12) we exclude v_3 by using (11) and obtain a quadratic equation for value v_2, solving which we find relationship variables v_1 and v_2 in equilibrium positions $\left(v_2\right)_{1,2} = \frac{1 - 24v_1 \pm \left(1 - 48v_1\right)}{12}$, i.e.

$$\left(v_2\right)_1 = \frac{1}{6} - 6v_1, \tag{14}$$

$$\left(v_2\right)_2 = 2v_1. \tag{15}$$

Substituting (14) into (13), can to obtain

$$\left(v_3\right)_1 = \frac{1}{36}. \quad (16)$$

Substituting (14), (16) into (11), can to obtain a quadratic equation for value $\left(v_1\right)_1$. Its decisions are $\left(v_1\right)_{1(1,2)} = \frac{3 \pm 1/3}{144}$, *i.e.* $\left(v_1\right)_{1(1)} = \frac{5}{216}, \left(v_1\right)_{1(2)} = \frac{1}{54}$.

With regard (14) we have $\left(v_2\right)_{1(1)} = \frac{1}{36}, \left(v_2\right)_{1(2)} = \frac{1}{18}$. Thus, we have the coordinates a two points

$$\left(v_1\right)_{1(1)} = \frac{5}{216}; \left(v_2\right)_{1(1)} = \frac{1}{36}; \left(v_3\right)_{1(1)} = \frac{1}{36}; \quad (17)$$

$$\left(v_1\right)_{1(2)} = \frac{1}{54}; \left(v_2\right)_{1(2)} = \frac{1}{18}; \left(v_3\right)_{1(2)} = \frac{1}{36}. \quad (18)$$

Substituting (15) into (13) can to obtain

$$\left(v_3\right)_2 = \frac{1}{12} - \frac{8}{3} v_1. \quad (19)$$

Substituting (15), (19) into (11) can to obtain a quadratic equation for value $\left(v_1\right)_2$. Its decisions are

$$\left(v_1\right)_{2(1,2)} = \frac{25 \pm 3}{1232}, \textit{i.e.} \left(v_1\right)_{2(1)} = \frac{1}{44}, \left(v_1\right)_{2(2)} = \frac{1}{56}.$$

With regard (15) its decision are $\left(v_2\right)_{2(1)} = \frac{1}{22}, \left(v_2\right)_{2(2)} = \frac{1}{28}$.

Thus, with regard (19) the coordinates yet two points are

$$\left(v_1\right)_{2(1)} = \frac{1}{44}; \left(v_2\right)_{2(1)} = \frac{1}{22}; \left(v_3\right)_{2(1)} = \frac{1}{44}; \quad (20)$$

$$\left(v_1\right)_{2(2)} = \frac{1}{56}; \left(v_2\right)_{2(2)} = \frac{1}{28}; \left(v_3\right)_{2(2)} = \frac{1}{28}. \qquad (21)$$

Let`s designate the point defined by equation (17) as point 5); point defined by (20) as point 6); point defined by (21) as point 7); point defined by (18) as point 8).

Linearizing the system of equations (8) - (10) in the vicinity of each of the 8 fixed points we can find the eigenvalues of the equilibrium positions. Distributing in a Taylor series right sides of equations (8) - (10) on three phase variables in the vicinity of the equilibrium positions, restricting oneself to a linear deviations from the equilibrium position, from the received system discriminant for points 1) - 4), we find the eigenvalues of the equilibrium positions. At point 1) eigenvalues are $\lambda_1 = \lambda_2 = \lambda_3 = 1$; at point 2) $\lambda_1 = \frac{1}{5}, \lambda_2 = -\frac{1}{5}, \lambda_3 = -\frac{2}{5}$; at point 3) $\lambda_1 = 0, \lambda_2 = \lambda_3 = 1$; at point 4) $\lambda_1 = \frac{1}{3}, \lambda_2 = \frac{\sqrt{97}-1}{12}, \lambda_3 = -\frac{\sqrt{97}+1}{12}$. Thus, the equilibrium position 1) - 4) in three - dimensional phase space unstable, since each of them has positive eigenvalues. Types of three - dimensional singular points were determined by classification (Blaquiere, 1966). Point 1) is unstable node, point 2) is saddle - node, point 3) is a degenerate unstable node, point 4) is saddle - node. At points 5) - 8) discriminant in the vicinity of singular points lead to the third - order polynomials, which were solved numerically.

For point 5) a third - order polynomial has the form

$$\lambda^3 + 0,166666\lambda^2 - 0,2407499\lambda + 0,0233196 = 0,$$

its decision are $\lambda_1 = 0,1; \lambda_2 = 0,261; \lambda_3 = -0,6.$

For point 6) third - order polynomial has the form

$$\lambda^3 + 0,289256\lambda - 0,02854996 = 0,$$

its decision $\lambda_1 = 0,09; \lambda_{2,3} = -0,0493506 \pm i0,54457566.$

Forpoint 7) third - order polynomial has the form

$$\lambda^3 + 0,78571428\lambda^2 - 0,540816326\lambda + 0,12536443 = 0,$$

its decision are $\lambda_{1,2} = 0,2471428 \pm i0,175470216; \lambda_3 = -1,28.$

For point 8) third - order polynomial has the form

$$\lambda^3 + 0,4074074\lambda^2 - 0,50617284\lambda - 49 = 0,$$

its decision are $\lambda_1 = 3,3; \lambda_{2,3} = -1,8537037 \pm i2,87959.$

Thus, all the fixed points 5) - 8) are also unstable, as each of these points among the eigenvalues has positive eigenvalues, or they have a positive real part. Equilibrium position 5) has the type of unstable saddle; equilibrium position 6) is of the type saddle - focus; equilibrium position 7) is of the type unstable saddle - focus; equilibrium position 8) is of the type saddle - focus.

Consequently, all the equilibrium position of system (8) - (10) are unstable. This fact contains a fundamental distinction from the conclusions of (Mukamel & Krinsky, 1976b), in which as a result of incorrect identification of eigenvalues of equilibrium positions the existence of stable equilibrium positions of the system among the states of equilibrium of system (8) - (10) was asserted.

To determine the character of behavior of solutions of the system (8) - (10) not only in the vicinity of equilibrium in the linear approximation, but also in the phase space as a whole, let`s construct the zero surface of the system (8) - (10) and examine their lines of intersection. Equation of zero surface F_1 is the equation of the surface shown in Figure 5.

Within the region bounded by the surface F_1 and coordinate planes $v_2 = 0, v_3 = 0,$ the variable v_1 increases in the outer region it reduces. This indicate the arrow at zero surface. Zero surface F_2 is shown in Figure 6.

Within the region bounded by the surface F_2 variable v_2 increases, out of this region it falls.

Zero surface F_3 is a plane (Figure 7) intercept on axes v_1, v_2, v_3 segments, 1/24, ¼, 1/12 respectively. Under the plane variable v_3 increases, above the plane it reduces.

Intersection of zero surfaces of the system (8) – (10) is shown in Figure 8.

General points of all four surfaces are equilibrium positions of system (fixed points 1) - 8)). Using data from the analysis of the equilibrium positions (their coordinates and eigenvalues), mutual position of zero surfaces forming region in the phase space regions with permanent derivative signs from phase coordinates, one can imagine the set of trajectories in phase space. It should

Figure 5. Zero surface F_1 of system (8) – (10)

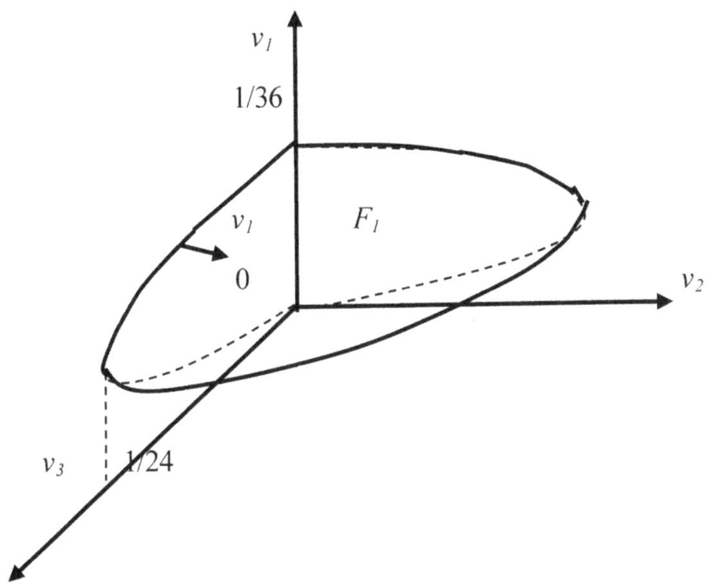

Figure 6. Zero surface F_2 of system (8) – (10)

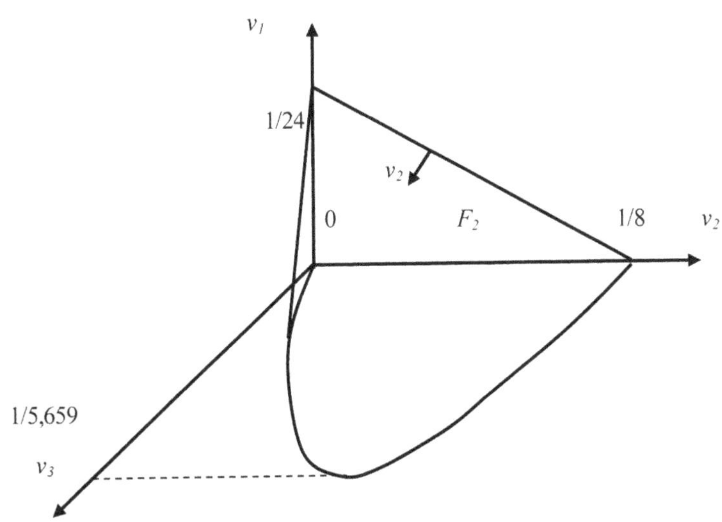

Figure 7. Zero surface F_3 of system (8) – (10)

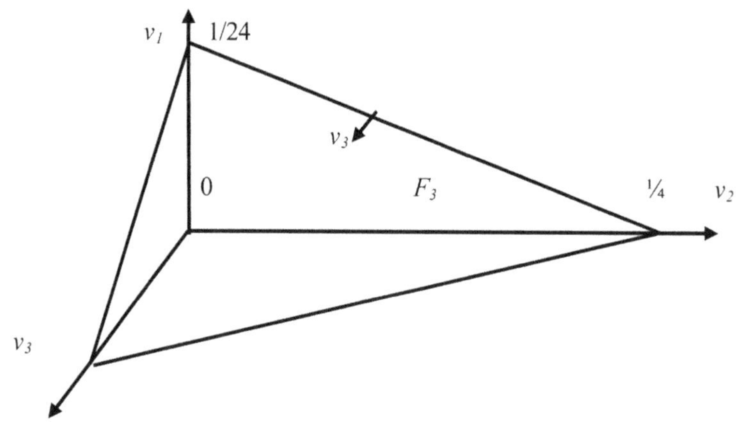

Figure 8. Intersection of zero surfaces of system (8) – (10)

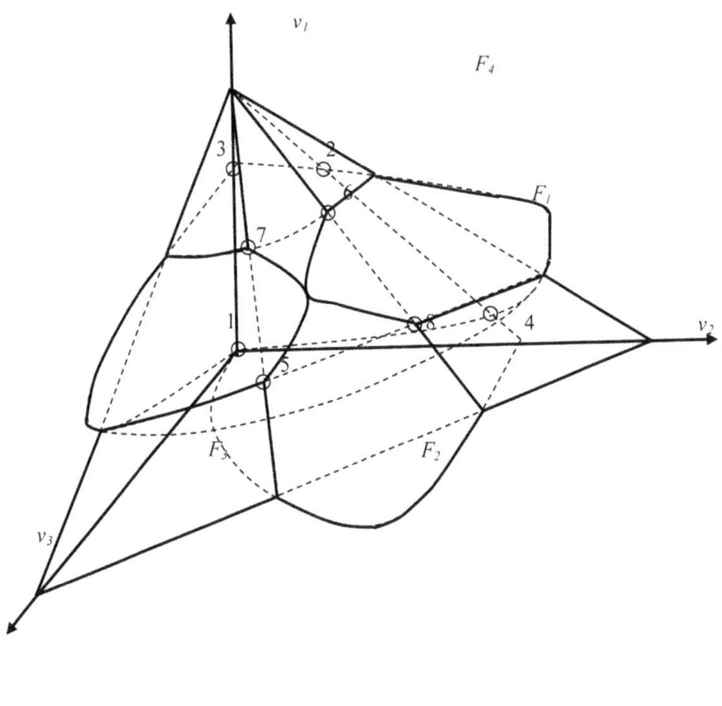

be noted that the zero surface F_4 is at the same time an integral plane of the system. To simplify the three-dimensional phase portrait of the system, many trajectories on the plane $v_3 = 0$ it is shown separately in Figure 9, and the system trajectories in three-dimensional phase space are shown in Figure 10.

Figure 9. Integral surface of system (8) – (10)

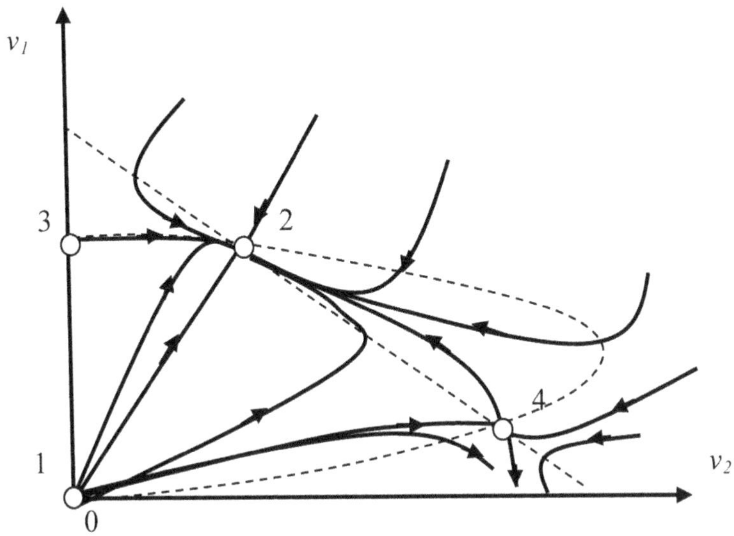

Figure 10. Phase portrait of system (8) – (10)

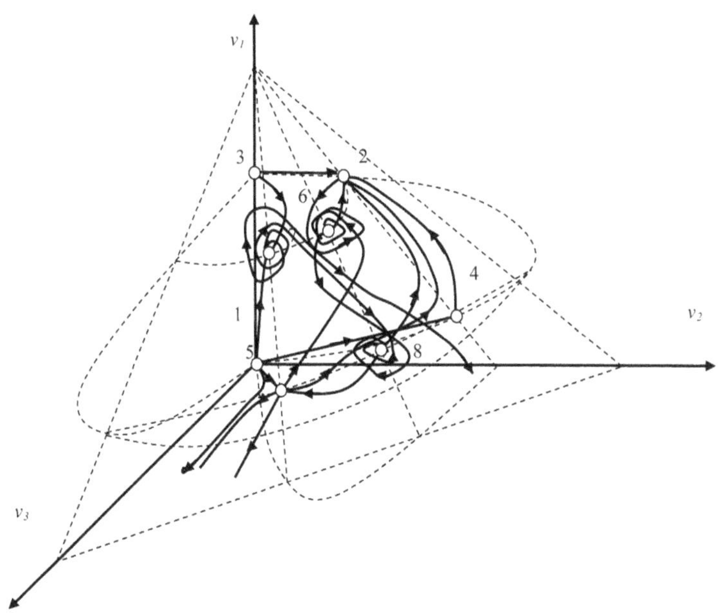

It is evident that the absence of stable equilibrium positions in the system under the included in the system of renormalization group equations condition of reducing of values of all phase coordinates at the sufficient distance from the beginning of coordinates outside zero surfaces results in chaotic transitions between equilibrium positions 2), 4) - 8), accompanied by fluctuation of phase coordinates. Since every of equilibrium position corresponds to a specific form of the Hamiltonian and, therefore, the geometric structure of a certain symmetry (Mukamel, 1975; Mukamel & Krinsky, 1976a, b), can to say that at high dimensionality of order parameter the phase transition is accompanied by fluctuation of the geometric structure.

Thus, the qualitative study of differential equations of the renormalization group of phase transitions with the order parameter higher dimensionality in multidimensional phase spaces showed that in the case of the six - dimensional order parameter all the equilibrium positions of the corresponding three - dimensional phase space are unstable. It means that the phase transition in these cases cannot finish at any particular geometric structure (a certain shape and a certain size of the unit cell). They change continuously, i.e. there is fluctuation of geometric structure - scaling fluctuation (Zhizhin, 2014a, b, c) takes place.

This received result fundamentally differs from the conclusions of (Mukamel & Krinsky, 1976b), which stated that the stable equilibrium positions in the studied system with a six -dimensional order parameter exist. The reason for this difference in the results is mathematical incorrectness of research in (Mukamel & Krinsky, 1976b). In particular, the coordinates of equilibrium positions in (Mukamel & Krinsky, 1976b) are in the linear approximation at ε, and the stability is defined taking into account the terms proportional to ε^2.

Inconsistency is hidden in the ε - resolution itself. If at first assumed that $\varepsilon << 1$, then, first of all, the quadratic terms of ε cannot lead to a significant change in the results. If there are changes, it indicates "structural instability" of the resulting system, which is to be worked out to the original rough system for any small "stirring" of the parameters values. Secondly, it is incorrect to believe that $\varepsilon = 1$ in the final formulas, if at the beginning of studies was suggested that $\varepsilon << 1$. We should stay within the limits of initial conditions, i.e. to consider the dimension of the space close to 4.

THE POLYTOPES OF HIGHER DIMENSION IN QUASI - CRYSTALS

In the theory of second - order phase transitions of Landau, Kadanov, Wilson, and Fisher (Landau, 1937; Kadanoff, 1966; Wilson, 1971; Fisher, 1965), elementary cells in a continuous condensed medium are considered which, in the course of a second - order phase transition, change their size, i.e., the scaling process is accompanied by a second - order phase transition. However, in this case, the specific form of elementary cells is not considered. Since the analysis shows that the process of phase transition occurs in a space with a dimension greater than three, it should be assumed that these elementary cells have the form of polytopes of higher dimension. But, what they are, it is impossible to establish in the framework of the theory of phase transition of the second kind. This requires experimental studies of the phase structure. For this purpose, it is possible to use diffraction patterns of intermetallic compounds, which are formed during the thermal treatment of metals accompanied by a change in the structure of metals, i.e. phase transition of the second kind. Investigations of the diffraction patterns of intermetallic led to the discovery of quasi - crystals.

Golden Hyper – Rhombohedron – Translational Basis of Quasi - Crystals

Finding in 1982 of ordered structures deprived (as it seemed) of translational symmetry (Shechtman et al., 1984), next called “quasi - crystals”, had marked the beginning of numerous cycles of papers and books devoted to the experimental and theoretical study of these unusual materials. Later it was found that the diffraction patterns of quasicrystals have a latent periodicity (Zhizhin, 2014), if we consider the diffraction pattern as a projection of a structure from a space of higher dimension. Figure 11 shows a typical diffraction pattern of intermetallic compounds.

Similar structures have other intermetallic compounds involving transition metals, for example: Al_6Mn (Shechtman et al., 1984), $Al_{70}Fe_{20}W_{10}$ (Mukhopadhyay et al., 1993), $Ti_{54}Zr_{26}Ni_{20}$ (Zhang & Kelton, 1993).

Figure 11 shows that the luminous points, which are a reflection of light from the impact of the electron beam, form five families of parallel lines oriented with respect to each other at angles of a multiple of 72 degrees. The distances between the parallel lines and the angles are determined by the

Figure 11. Electron diffraction pattern of compound $Al_{72}Ni_{20}Co_8$ (Eiji Abe et al., 2004)

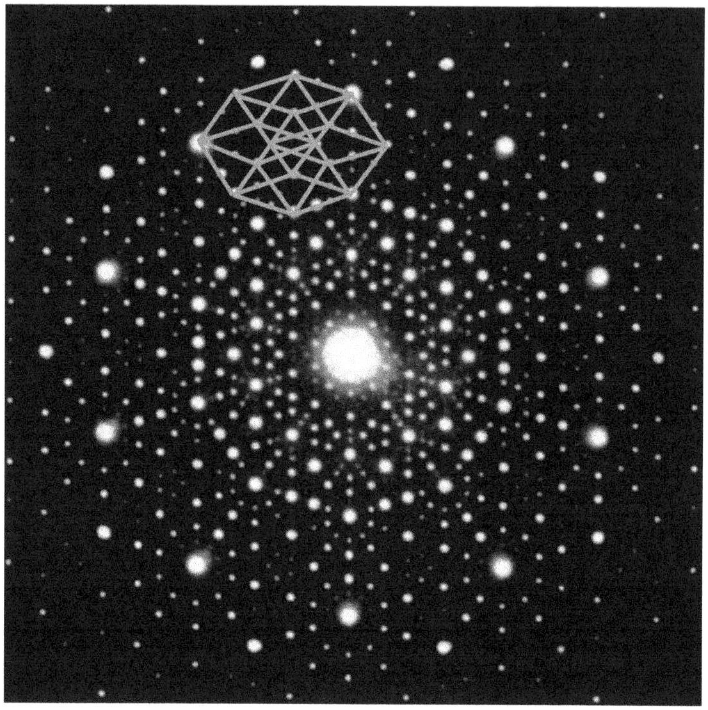

golden section. A geometrical model of the structure of the diffraction patterns of quasi - crystals was constructed (Shevchenko, Zhizhin & Mackey, 2013a; Shevchenko, Zhizhin & Mackey, 2013b; Zhizhin, 2014; Zhizhin & Diudea, 2016). It was shown that the elementary cell of this geometric structure is a polytope of dimension 4, which was called a gold hyper - rhombohedron. This cell is plotted on the diffraction pattern in Figure 11 with solid segments of light lines. I can see that it passes through the luminous points of the diffraction pattern observing its geometry. This cell fills the entire space with the translation reflected by the diffraction pattern. To determine the regularities of this cell, it is depicted in Figure 12 on an enlarged scale.

The polytope in Figure 12 consists of 8 three-dimensional polyhedrons whose faces are determined by construction with a golden section. Namely, all two-dimensional faces represent the same diamonds with angles $\varphi = \arccos\frac{1}{2\tau}, \gamma = \pi - \varphi$, where τ is golden section. The three -dimensional faces (rhombohedra) are adjacent to each other along the flat faces, so that all planar faces are common to some two rhombohedrons, which is necessary

Figure 12. Golden hyper-rhombohedron

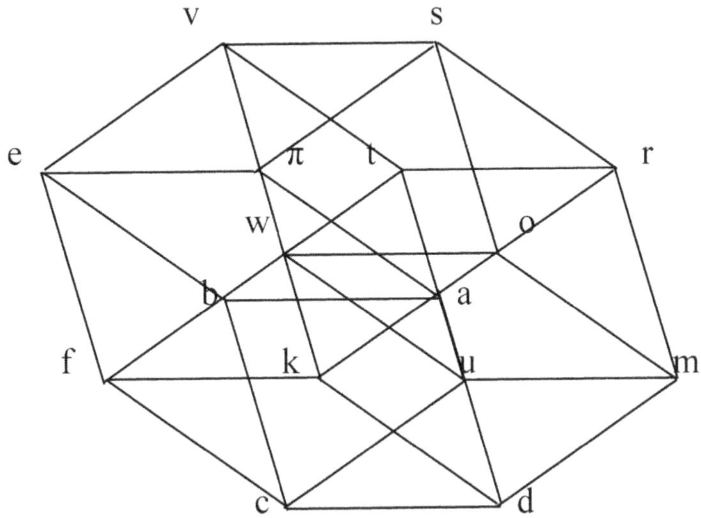

for the polytope to exist. The polytope in Figure 12 has the same number of vertices (16), edges (32), planar (24) and three -dimensional (8) faces as 4 - cub. Therefore, substituting these quantities of elements of different dimensions into the Euler - Poincaré equation (22), can get that this equation is satisfied for a value of dimension equal to 4. The difference from the 4 - cube is that its flat faces are rhombs, and not squares. Therefore, the golden hyper - rhombohedron is a semi - regular polytope, since its two - dimensional faces are not regular polygons, although all vertices are compatible with motion. The periodic part of the structure of quasi - crystals is the translational filling of a four - dimensional space by a golden hyper - rhombohedron. The edges emanating from its vertices carry out the product of the rhombohedron into a segment. Just like a 4 - cube is obtained by multiplying the cube by an edge. As a result, the number of elements of different dimensions in the gold hyper - rhombohedron is the same as in the 4 - cube. The difference from the 4 - cube lies in the fact that all its metric relationships are based on the golden section. It was shown (Zhizhin, 2014, Zhizhin& Diudea, 2016) that the lattice of the vertices of the golden hyper - rhombohedron during its translation highlights the latent periodicity of the diffraction patterns.

Polytopes of Various Dimensions and Shapes on the Lattice of Vertices of the Golden Hyper - Rhombohedron

On this lattice can be built all the Platonic bodies, the Bravais (Bravais, 1848) and Delone (Delone, Padurov& Alecsandrov, 1934; Delone, 1937) cells, Voronoi (Voronoi, 1908) cells. Figure 13 representing the lattice of golden hyper - rhombohedron vertices, as an example, shows how from the vertices of the golden hyper - rhombohedron one can obtain projections of Plato's bodies: tetrahedron, cube, octahedron, icosahedron, dodecahedron.

Moreover, these figures can vary in size. It is essential that in addition to the already built golden hyper - rhombohedron of certain size, these polytopes of other dimensions can be built on vertices lattice (see Figure 14).

On the Figure 13 and Figure 14 the vertices of golden hyper - rhombohedron big size is indicated of a white circles. On Figure 14 besides golden hyper - rhombohedron little size is indicated of a black segments.

The physical nature of the process of scaling in phase transitions of the second kind and in quasicrystals is a change in the structure during the phase transition of the second kind (Kadanoff, 1966), and in the processes of formation of quasi - crystals during annealing. However, if Kadanoff assumed for phase transitions the merging of smaller cells to larger ones, here the analysis of golden hyper - rhombohedron vertices lattice shows that the process of scale changing occurs hierarchically, i.e. a smaller hyper - rhombohedron cell is not combined with similar ones but simply increases in hierarchical mode, absorbing more and more of lattice sites. On the diffraction pattern (Figure 14), as an example, the projection of golden hyper - rhombohedron with

Figure 13. Plato`s bodies on the lattice of golden hyper - rhombohedron

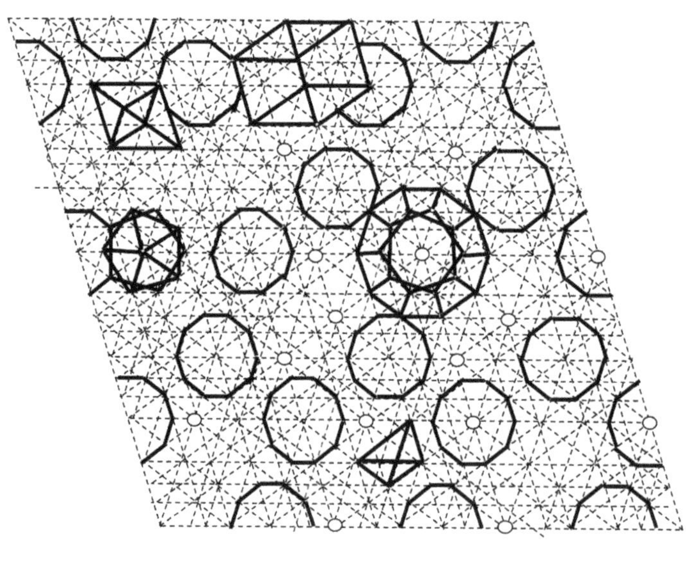

Figure 14. Golden hyper - rhombohedrons of different scale on the lattice of golden hyper -rhombohedron vertices

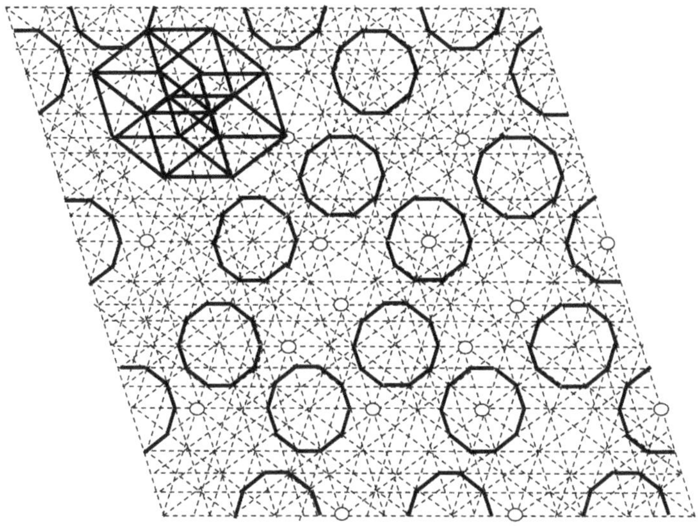

the edge of a certain size is applied. It's clear that hyper - rhombohedrons of other sizes can be distinguished on the diffraction pattern. The analysis of diffraction pattern geometry shows that the similarity coefficient of the hyper - rhombohedrons is the golden section τ. Raising τ in positive and negative integer powers one can receive the hyper - rhombohedrons of appropriate sizes, which can be identified on the diffraction pattern in the range of its solvability. It is essential that the coordinates of the vertices of the golden hyper - rhombohedron are expressed in forth - dimensional space in whole numbers. Let's choose the top *a* in Figure 12 as the beginning of coordinates system *(x, y, z, h)*. Let's direct the axis *x* along the edge *ar,* the axis *y* along the edge *aπ*, the axis *z* along the edge *ad,* the axis *t* along the edge *ab*. Let's assume that the length of hyper - rhombohedron edge is a unit. Then the coordinates of the vertices of the hyper - rhombohedron *aπebdkfc* are expressed in whole numbers *a(0, 0, 0, 0), π(0, 1, 0, 0), e(0, 1, 0, 1), b(0, 0, 0, 1), k(0, 1, 1, 0), f(0, 1, 1, 1), c(0, 0, 1, 1), d(0, 0, 1, 0)*. Now let's to change the coordinate x from 0 to 1. Then the coordinates of a*πebdkfc* hyper - rhombohedron vertices also to change v*(1, 1, 0, 1), t(0, 0, 0, 1),w(1, 1, 1, 1), u(1, 0, 1, 1), s(1, 1, 0, 0), r(1, 0, 0, 0), o(1, 1, 1, 0), m(1, 0, 1, 0)*. It's obvious that at hyper - rhombohedron translation the coordinates of all lattice vertices formed by hyper - rhombohedron vertices are integers. Filling the space of the hyper - rhombohedrons of other sizes also forms integral lattice, if the size of hyper - rhombohedron edges is mistaken for one. It's interesting that on the lattice of golden hyper - rhombohedron vertices in the space 4D it is possible to construct the projections of the golden rhombohedron from a space of even more dimension. Figure 15 shows a projection of the golden rhombohedron from the space 5D on the plane 2D.

It is clear that this projection can be laid on the lattice of golden hyper - rhombohedron vertices in the space *4D* in Figure 13. The 5 – rhombohedron is the product of hyper - rhombohedron by one-dimension segment. If length of the segment equal of length edges of the hyper - rhombohedron, so the 5-rhombohedron is semi - regular polytope too. The 5 - rhombohedron can also fill the space of quasi - crystal displayed by the diffraction pattern in Figure 11, i.e. one can say that the quasi - crystal has a subspace of dimension 5. Similarly, a rhombohedron of even more dimension can be built on the diffraction pattern. This leads to the conclusion that quasi-crystals have subspaces of any integer dimension.

From the said above it follows that the lattice of the vertices of the golden hyper - rhombohedrons have the universal nature. On it you can find the projections of the polytope 4 - simplex and other polytopes of higher dimension.

Figure 15. The golden rhombohedron in space 5D

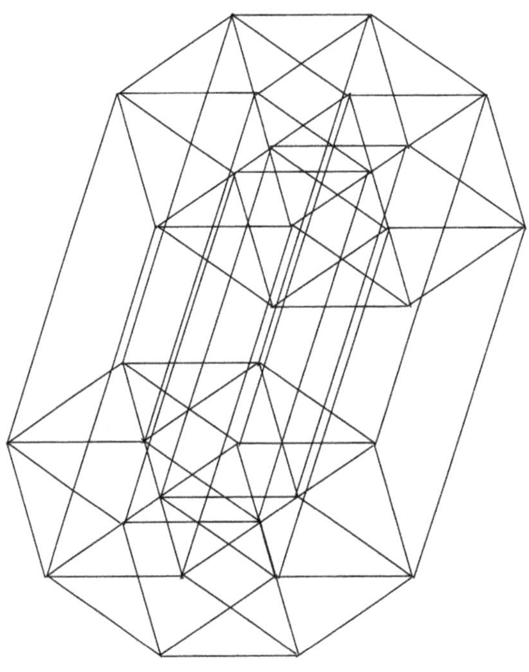

It is also important to note that on the diffraction patterns of intermetallic (Figure 11) one can see not only the latent periodicity of the structure in spaces of higher dimension, but also the hierarchical filling of spaces in the vicinity of almost any point of the structure. It is expressed in an increase in the size of the figures around an arbitrary point of the structure while maintaining the shape of the figure.

THE STRUCTURE OF DNA TAKING INTO ACCOUNT THE HIGHER DIMENSION OF ITS COMPONENTS

From the History Opening of the Deoxyribose Nucleic Acid (DNA)

DNA, as a chemical substance, it was isolated by Johann Friedrich Micher in 1869 from the remains of cells contained in pus. He singled out a substance that includes nitrogen and phosphorus. When Misher determined that this substance has acid properties, the substance was called nucleic acid (Dahm, 2005).

Gradually it was proved that it was DNA, and not proteins, as previously thought, and which is the carrier of genetic information. One of the first decisive proofs was the experiments of Oswald Avery, Colin MacLeod and McLean McCarthy (1944) on the transformation of bacteria. They managed to show that the DNA isolated from the pneumococci corresponds to the so -called transformation (the acquisition of pathogenic properties by a harmless culture as result of the addition of dead pathogenic bacteria to it). The experiment of American scientists Alfred Hershey and Martha Chase (Hershey - Chase experiment, 1952) with radioactively labeled proteins and bacteriophage DNA showed that only the phage nucleic acid is transmitted to the infected cell, and the new generation of phage contains the same proteins and nucleic acid, as the initial phage (Hershey & Chase, 1952). Deciphering the structure of DNA (1953) has become one of the turning points in the history of biology. For outstanding contributions to this discovery, Francis Crick, James Watson and Maurice Wilkins were awarded the 1962 Nobel Prize in Physiology or Medicine (Westly, 2008).

The structure of the double helix DNA it was proposed by Francis Crick and James Watson in 1953 on the base of the X - ray structural data obtained by Maurice Wilkins and Rosalind Franklin and the "Chargaff rules" according to which in each DNA molecule the strict relationships connecting the quantity of nitrogenous bases of different types (Watson & Crick, 1953). In 1986, Frank - Kamenetskii in Moscow showed how a double-stranded DNA folds into a so-called H - shape, composed not of two but three strands of DNA (Frank – Kamenetskii, 1986, 1988). Deoxyribonucleic acid (DNA) is a biopolymer, the monomer of which is the nucleotide (Albert et al., 2002; Butler, 2001).

Each nucleotide consists of a phosphoric acid residue attached to sugar deoxyribose, to which one of the four nitrogen bases is attached also. The bases that make up the nucleotides there is divided into two groups: purines (adenine [A] and guanine [G]) are formed by combined five - and six - membered heterocycles; pyrimidines (cytosine [C] and thymine [T]) – six - membered heterocycles.

Nucleotides are long polynucleotide chains covalently linked. These chains in the overwhelming majority of cases (except for some viruses possessing single - stranded DNA genomes) there are combined pairwise by means of hydrogen bonds into a secondary structure, called the double helix (Watson & Crick, 1953; Berg, Tymoczko & Stryer, 2002). Each base on one of the chains is connected to one definite base on the second chain. This specific binding is called complementary. Purines are complementary to pyrimidines (that is, they are capable of forming hydrogen bonds with them): adenine forms bonds only with thymine, and cytosine - with guanine. In a double helix, chains are also linked by hydrophobic interactions and stacking, which do not depend on the DNA base sequence (Ponnuswamy & Gromiha, 1994). Complementarity of the double helix means that the information contained in one chain is also contained in another chain. Complementarity of the double helix means that the information is contained in one chain. Different base pairs form a different number of hydrogen bonds.

Watson and Crick postulated the spiral form of the DNA molecule, but they did not discuss the reasons for the formation of such a DNA molecule. Until now there have been no works explaining the existence of a spiral in the DNA molecule. In recent works of the author (Zhizhin, 2016b, 2017) it was shown that many molecules, including biomolecules, have higher dimensionality. Therefore, it is of interest to consider the question of the formation of the spiral structure of the DNA molecule, taking into account the higher dimensionality of its components.

The Dimension of Phosphoric Acid and Its Residue

The phosphoric acid H_3PO_4 has the following building (see, for example, Karapetians & Drakin, 1994) (Figure 16).

It is obvious that, used in the DNA molecule, the phosphoric acid residue obtained by detaching one hydrogen atom from it has the building also. In this case, instead of one of the hydroxyl groups attached to the phosphorus atom, there will be a negatively charged oxygen atom. To determine the functional topological dimension of phosphoric acid and its residue, it is necessary to connect the vertices, atoms and functional groups entering into these compounds (Zhizhin, 2016b, 2017). Then the structure of phosphoric acid and its residue is represented in the form of a polytope in Figure 17.

Calculation of the functional topological dimension by the Euler - Poincaré equation (Poincaré, 1895) does not require specifying the exact lengths of polytope edges, and for a given number of the vertices does not depend on the groups of atoms in these vertices (Zhizhin, 2016b). In Figure 17, the edges corresponding to the covalent chemical bonds are marked in red, and the edges that carry only the geometric function, delineating the closed geometric figure, are marked in blue.

It follows from Figure 17 that the polytope of phosphoric acid has five vertices (*a, b, c, d, e*), ten edges(*ac, ab, ad, ae, cd, de, be, ce, cb, bd*), ten plane triangular faces (*cbe, cda, cea, dab, dbe, dce, bae, dae, bce, dcb*), and five three - dimensional faces in the form of tetrahedron (*cbae, caed, cbad, abde,*

Figure 16. The building of the phosphoric acid

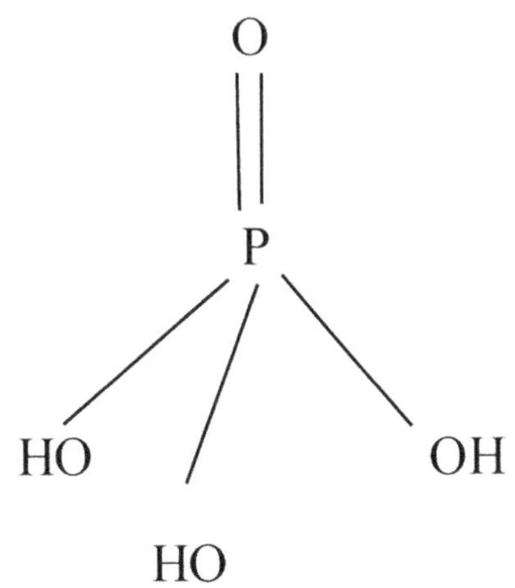

Figure 17. The polytope of the phosphoric acid and its residue

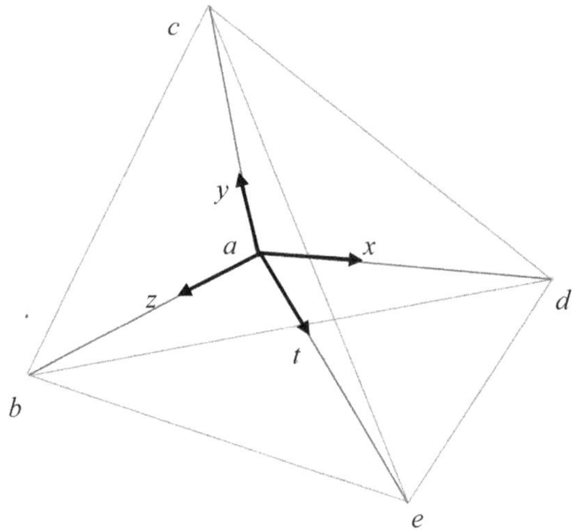

cebd). These values of elements of different dimensions we must substitute into the Euler - Poincare equation

$$\sum_{i=0}^{n-1} f_i \left(-1\right)^i = 1 - \left(-1\right)^i \tag{22}$$

where f_i is the number of elements with dimension *i* in polytope with dimension *n*. In this case there are $f_0 = 5, f_1 = 10, f_2 = 10, f_3 = 5$.

Substituting these values into equation (22), we obtain

$5 - 10 + 10 - 5 = 0.$

Thus, equation (22) is satisfied for *n* = 4. This proves that the polytope of phosphoric acid and its residue has a dimension of four. It is worth emphasizing that the polytope of phosphoric acid and its residue is topologically equivalent to a convex polytope of the simplex type with dimension four, since these polytopes completely coincide with the values of the numbers of elements of different dimensions (Zhizhin, 2014b). The origin of the four-dimensional space can be placed in one of the vertices with sending coordinates along the edges connecting this vertex to other vertices (see Figure 17).

The Dimension of the Deoxyribose

The deoxyribose (or the *D* – ribose) is molecule of carbohydrate with five atoms carbon. Its image in the form of Fisher's formula (Metzler, 1980) is shown in Figure 18.

In the DNA molecule, *D* - ribose as a component enters with a closed carbon chain. The closed chain of the molecule in Figure 18 is formed due to the rupture of the double bond between oxygen and carbon (considering it the first in the chain), the liberation of the water molecule, and the addition of this oxygen atom to the penultimate (fourth) carbon atom in the carbon chain. The Fischer formula or formula Haworth and their modifications (Metzler, 1980; Lehninger, 1982) may not reflect the spatial structure of the *D* – ribose molecule. For this target the constructs described in the form of convex polytopes with boundary elements which form boundary complex (Grunbaum, 1967). Only when such a representation will be to determine the dimension of these molecules (Zhizhin, 2016b; 2017). Figure 19 shows the result of the closure of a chain of four carbon atoms through an oxygen atom. Chemical bonds are marked as before with red color.

It is easy to see that, in addition to the form of the *D* - ribose molecule in Figure 19 (a form of *A*), there is another enantiomer form of this molecule. It connected by a symmetry transformation with respect to a straight line

Figure 18. The molecule D – ribose (formula of Fisher's)

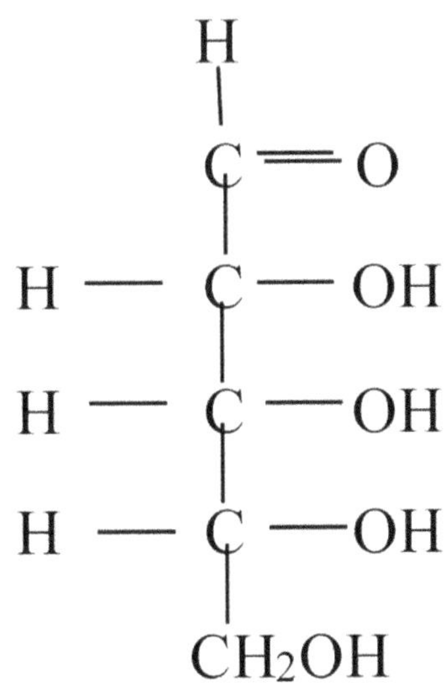

Figure 19. The molecule D – ribose with closed a chain of carbon atoms (the form A)

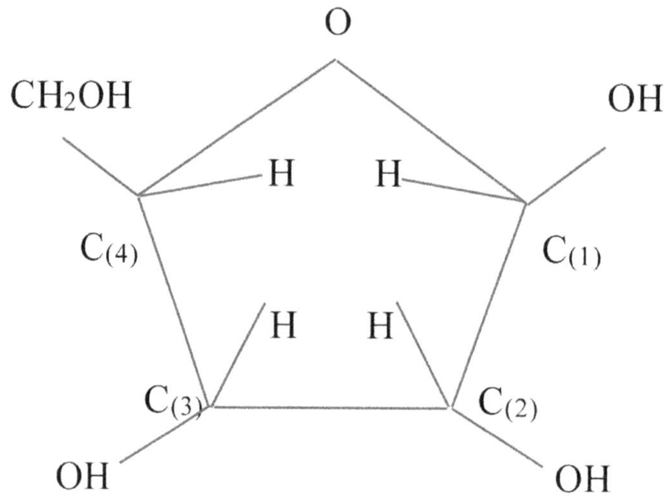

passing through an oxygen atom separating the projection of the form of *A* in half (we shall assume that it is a *B* shape). Form *B* of the *D* - ribose molecule is shown in Figure 20.

In order to give the *D* - ribose molecule a kind of polytope, it is necessary to connect each vertex (every atom or functional group, with all other vertices) to the ribs, then we get a polytope of the simplex type, the dimension of which must be determined.

Figure 21 shows an image of a *D* - ribose molecule (*A*) with a closed carbon chain in the form of a polytope.

In contrast to the previous works of the author (Zhizhin, 2016b; 2017), all hydroxyl groups are transferred to the periphery of the image of the molecule. This is more convenient for considering possible chemical reactions of *D* - ribose with other compounds.

The polytope at Figure 21 has 13 vertices in which there are one oxygen atom, four hydrogen atoms, three hydroxyl groups, four carbon atoms, compound CH_2OH. It is known that the dimension of a simplex is one less than the number of its vertices (Grünbaum, 1967). Therefore, the dimension of the polytope in Figure 21, if the Euler - Poincaré equation (22) holds for it, equals $n = f_0 - 1 = 12$. The simplex is a simplicial polytope, and all its faces therefore are simplexes. Since all the vertices of a simplex are related to each other, the number of edges is determined by the number of combinations of the number of vertices of two. In this case, the number of edges of the

Figure 20. The molecule D – ribose with closed a chain of carbon atoms (the form B)

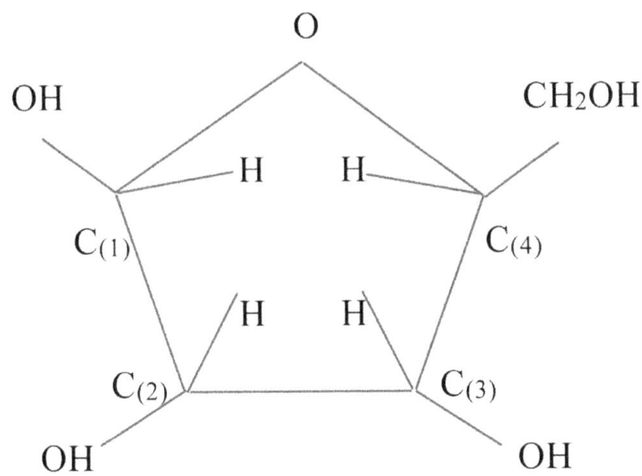

Figure 21. The molecule D – ribose (A) in form of the polytope

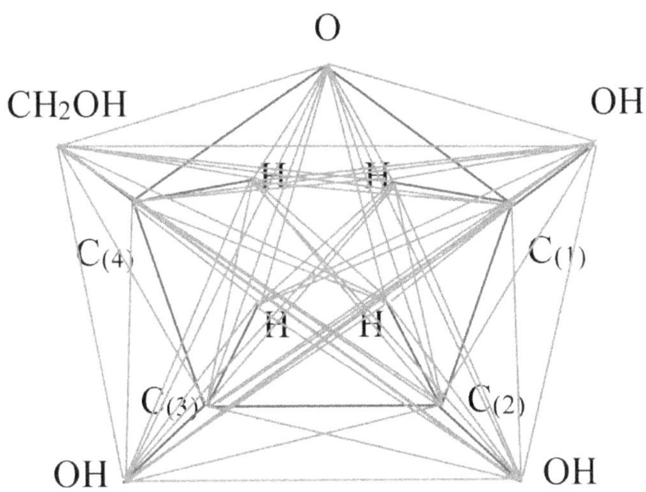

polytope in Figure 6 is $f_1 = C^2_{n+1} = 78$. The number of two - dimensional faces (triangles) of this polytope is $f_2 = C^3_{n+1} = 286$.

The number of three - dimensional faces (tetrahedrons) of this polytope is $f_3 = C^4_{n+1} = 715$.

The number of four – dimensional faces (4 – simplexes) of this polytope is $f_4 = C^5_{n+1} = 1287.$

The number of five – dimensional faces (5 – simplexes) of this polytope is $f_5 = C^6_{n+1} = 1716.$

The number of six – dimensional faces (6 – simplexes) of this polytope is $f_6 = C^7_{n+1} = 1716.$

The number of seven – dimensional faces (7 – simplexes) of this polytope is $f_7 = C^8_{n+1} = 1287.$

The number of eight – dimensional faces (8 – simplexes) of this polytope is $f_8 = C^9_{n+1} = 715.$ The number of nine – dimensional faces (9 – simplexes) of this polytope is $f_9 = C^{10}_{n+1} = 286.$

The number of ten – dimensional faces (10 – simplexes) of this polytope is $f_{10} = C^{11}_{n+1} = 78.$

The number of eleven – dimensional faces (11 – simplexes) of this polytope is $f_{11} = C^{12}_{n+1} = 13.$ Substituting the values $f_i, \left(0 \le i \le 11\right)$ in the Euler - Poincaré equation (22), we see that it holds for $n = 12$ for polytope on Figure 21

$$13 – 78 + 286 – 715 + 1287 – 1716 + 1716 – 1287 + 715 – 286 + 78 – 13 = 0.$$

This finally proves that the polytope *D* - ribose (*A*) has dimension 12 and a simplex type. Obviously, the polytope *D* - ribose (*B*) has the same dimension and type.

The Structure of the Deoxyribose Nucleic Acid

The DNA is a chain of *D* - ribose molecules between which are the phosphoric acid residues that connect these molecules. Let us construct this circuit, taking into account the concept of repulsion of electron pairs adopted in the scientific literature (Gillespie, 1972; Gillespie & Hargittai, 1991). According to this concept, tetrahedral coordination of atoms and functional groups in these compounds is carried out around the phosphorus atom in the phosphoric acid residue, and in the carbon atoms in the *D* - ribose molecule also. It is this concept that leads to the higher dimensionality of the remainder of phosphoric acid and the molecule of *D* - ribose. When a phosphoric acid residue is added to the *D* - ribose molecule from the hydroxyl group of the phosphoric acid residue and the hydroxyl group of the *D* - ribose molecule, a water molecule that leaves the compounds forms and an oxygen atom that binds the phosphoric acid residue and the *D* - ribose molecule. The concept of repulsion of electron pairs also operates here, according to which two valence electrons of the oxygen atom maximally repel each other, forming electron pairs with electrons of the phosphorus atom and the carbon atom. Therefore, taking into account that tetrahedral coordination is carried out around the atoms of phosphorus and carbon already, we can assume that the valence bonds of the oxygen atom are located linearly with respect to the oxygen atom. When the second *D* - ribose molecule is added to the phosphoric acid residue, the second hydroxyl group of the phosphoric acid residue with the compound CH_2OH of this *D* -ribose molecule also form molecules of water that leaves the compound and an oxygen atom connecting the phosphorus atom and the functional group CH_2 of the *D* - ribose molecule. By virtue of the concept of repulsion of electron pairs, the linear arrangement of the valence bonds of the oxygen atom can also be considered here. We will take this into account when constructing a chain of DNA molecules and phosphoric acid residues. To construct the sequence of *D* - ribose molecules and the phosphoric acid residue in the chain, we will use a simplified image of the *D* - ribose molecule of dimension 12. Let us leave only the outer contour and valence bonds necessary for interaction with the environment from the *D* - ribose molecule (Figure 6). At the same time, we will not depict the hydrogen atoms inside the contour of carbon atoms (Figure 22).

The absence of edges and atoms in Figure 22 in comparison with the figure does not mean that they really are not. This is only the conventionality of the image. Now, in a projection onto a two-dimensional plane, we shall depict a sequence of simplified images of *D* - ribose molecules and phosphoric acid residues, taking into account the separation of water molecules at the junction of phosphoric acid residues and *D* - ribose molecules and the linear arrangement of the valence bonds in vicinity of these compounds (Figure 23).

From Figure 23 it immediately follows that the molecules are not in a straight line. The dashed line along which the same hydroxyl groups of the *D* - ribose molecules are located has constant curvature. Taking into account that the phosphoric acid residue is not flat, this means that after passing from the previous *D* - ribose molecule to the subsequent *D* - ribose molecule through the phosphoric acid residue, the subsequent *D* - ribose molecule shifts to the right and upwards relative to the previous *D* - ribose molecule.

Figure 22. A simplified image of the D - ribose molecule

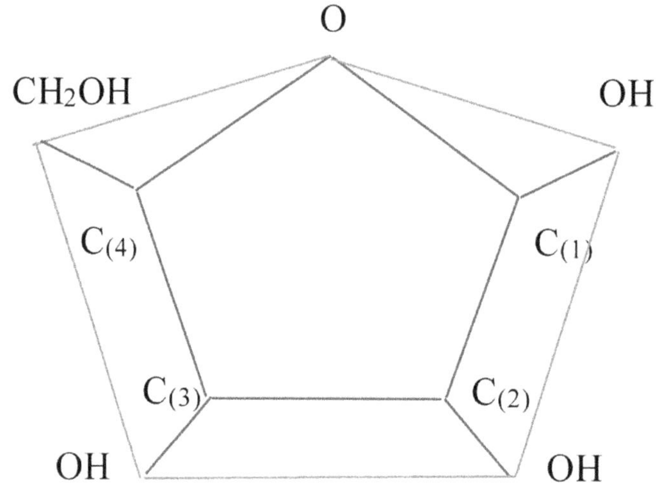

Figure 23. Structure of the DNA molecule

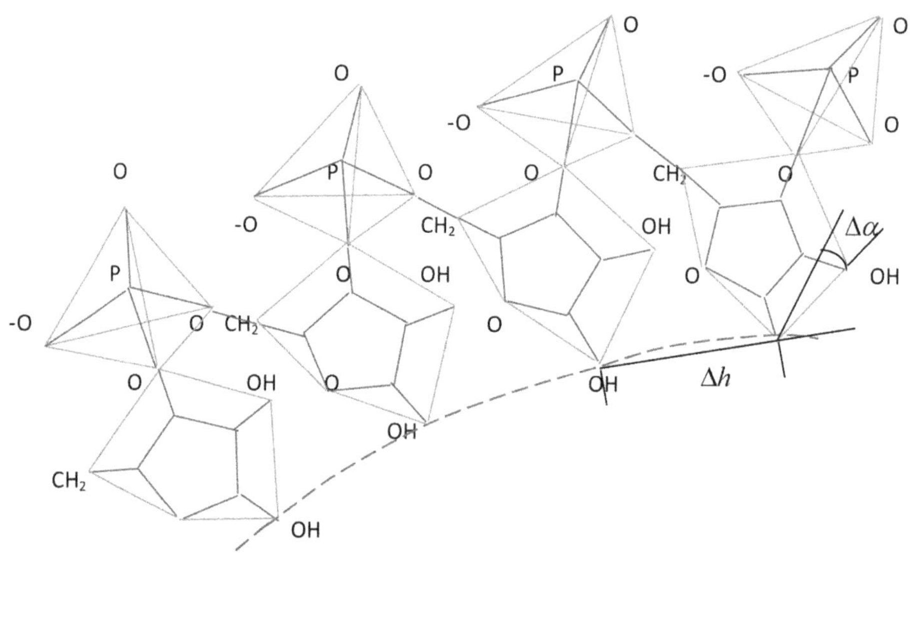

Thus, it follows from the geometric construction of the alternating sequence of *D* - ribose molecules and phosphoric acid residues, taking into account their dimensionality, that this sequence is located on the helical surface. This was exactly what Watson and Crick (Watson & Crick, 1953) postulated in their paper. Moreover, it follows from Figure 23 that this helical surface can be described mathematically accurately by the equation

$$\Delta h = k\left(l_{P-O}, l_{C-O}, l_{C-C}\right)\Delta\alpha \tag{23}$$

In the equation (23) it was used the following notation:

Δh is the smallest distance between two corresponding points of *D* - ribose molecules
$\Delta\alpha$ is the smallest angle between two corresponding edges of *D* – ribose molecules
k is the coefficient of proportionality
l_{P-O} is the length of chemical bond between of the phosphorus atom and the oxygen atom
l_{C-C} is the length of chemical bond between two the carbon atoms
l_{C-O} is the length of chemical bond between the carbon atom and the oxygen atom

In Figure 23, the form *B* of the *D* - ribose molecule it was used. If to use the form *A* of the *D* -ribose molecule, can to obtain also the arrangement of the sequence of molecules on the right of a twisted helical surface. However, in this case, the vertices with the oxygen atom will be located on the right side of the chain of carbon atoms. This is consistent with the depiction of the structure of two DNA molecules by Watson and Crick (Metzler, 1977). From the opposite of the phosphoric acid residues to the *D* – ribose molecules in the chain of molecules, they are joined by the hydrogen bonds of the base, which determine the genetic properties of the DNA. In Figure 23, these bases are not depicted, since they are of a flat nature, and this article draws attention to the higher dimensionality of molecules. If we consider the chain of molecules *D* - ribose and the residues of phosphoric acid from the opposite end of the chain, then naturally the helical surface will be twisted in the opposite direction to the left.

CONCLUSION

More Plato suggested that objects of Nature, their quality and operating between the forces, may be the result of exposure to hidden from us the geometric structures (Yau & Nadis, 2010). In recent decades, a shining example of the opening of the geometric structure is string theory. Based on the solutions of Einstein's equations, this theory postulated that the manifold of higher dimension hidden in every point in space (Colabi -Yau manifold of dimension 6). It has vanishingly small size. In the nanoworld, as shown in (Zhizhin, 2014a, b; Zhizhin, 2015 a, b; Zhizhin & Diudea, 2016; Zhizhin, Khalaj & Diudea, (2016); Zhizhin, 2016a, b) is also a hidden dimension but objects (especially molecules), having these dimensions, are tangible sizes. A common basis for understanding the hidden dimensions of the objects of nature, and in string theory, explains the structure of the universe and the structure of elementary particles (Green, 2000; Gepner, 1988), and in the theory of the nanoworld is Riemann geometry (Riemann, 1854), allowing the existence of spaces of higher dimension in the space of lower dimension.

The higher dimensionality of the constituent DNA molecules, which allows to describe mathematically the structure of DNA, requires reconsidering the issues of tight packing of DNA molecules in cells, viruses and bacteria, provided that the DNA chains necessary for the preservation and transfer of genetic information are complementary.

REFERENCES

Aharony, A. (1973). Critical Behavior of Anisotropic Cubic Systems. *Physical Review B: Condensed Matter and Materials Physics*, *8*(9), 4270–4273. doi:10.1103/PhysRevB.8.4270

Albert, B. (2002). *Molecular Biology of the Cell* (4th ed.). New York: Garland Science.

Anderson, J. E. (1963). *Magnetohydrodynamic shock waves*. Cambridge: M.I.T. Press.

Berg, J., Tymoczko, J., & Stryer, L. (2002). *Biochemistry*. W.H. Freeman & Co Ltd.

Blaquiere, A. (1966). *Nonlinear system analysis*. Paris: Academic Press.

Bogoyavlensci, O. I. (1980). *Methods of the qualitative theory of dynamical systems in astrophysics and gas dynamics*. Moscow: Science.

Braveis, A. (1848). *Abhandlung über die Systeme von regelmässing auf einer Ebene oder im Raum vertheilten Puntsten*. Leipzig: Ostwald`s Klassiker.

Butler, J. M. (2001). *Forensic DNA Typing*. Elsevier.

Coleman, E. (1970). *The Fourth Dimension*. Moscow: Science.

Dahm, R. (2005). Friedrich Miescher and the discovery of DNA. *Developmental Biology*, *278*(2), 274–288. doi:10.1016/j.ydbio.2004.11.028 PMID:15680349

Delone, B. (1937). Geometry of positive quadratic forms. *Uspekhi Matematicheskhi Nauk*, *3*, 16–62.

Delone, B., Padurov, N., & Aleksandrov, A. (1934). *Mathematical foundations of the structural analysis of crystals*. Leningrad, Moscow: ONTI.

Eiji, A., Yanfa, Y., & Pennycook, S. J. (2004). Quasicrystals as cluster aggregates. *Nature Materials*, *3*(11), 759–767. doi:10.1038/nmat1244 PMID:15516956

Einstain, A. (1930). Fundamental ideas and problems of the theory of relativity. *Die Koralle*, *5*, 486–487.

Euclid. (2012). *Beginnings*. Moscow: URSS.

Fisher, M. E. (1965). *The nature of critical points*. Colorado: Colorado University Press.

Fisher, M. E., & Pfeuty, P. (1972). Critical behavior of the anisotropic n – vector model. *Physical Review B: Condensed Matter and Materials Physics*, *6*(5), 1889–1891. doi:10.1103/PhysRevB.6.1889

Frank-Kamenetskii, M. (1986). DNA structure: A simple solution to the stability of the double helix? *Nature, 324*, 325.

Frank-Kamenetskii, M. (1988). H – form DNA and the hairpin – triplex model. *Nature, 333*, 214.

Gepner, D. (1988). Yukawa Couplings for Calabi-Yau String Compactification. *Nuclear Physics B*, *311*(1), 191–204. doi:10.1016/0550-3213(88)90147-2

Gillespie, R. J. (1972). *Molecular Geometry*. New York, Cincinnati, Toronto, Melbourne: Van Nostrand Reinhold Company.

Gillespie, R. J., & Hargittai, I. (1991). *The VSEPR Model of Molecular Geometry*. New York, Dover.

Green, B. (2000). *The Elegant Universe*. New York: Vintage Books.

Grunbaum, B. (1967). *Convex Polytopes*. London: Springer.

Hershey, A., & Chase, M. (1952). Independent functions of viral protein and nucleic acid in growth of bacteriophage. *The Journal of General Physiology*, *36*(1), 39–56. doi:10.1085/jgp.36.1.39 PMID:12981234

Izyumov, U. A., & Syromyatnikov, V. N. (1984). *Phase transitions and symmetry of crystals*. Moscow: Science.

Kadanoff, L. P. (1966). Scaling Laws For Ising Models Near τ_c^*. *Physics*, *2*(6), 263–272. doi:10.1103/PhysicsPhysiqueFizika.2.263

Karapet'yants, M. X., & Drakin, S. I. (1994). *General and Inorganic Chemistry*. Moscow: Chemistry.

Landau, L. D. (1937). To Theory of Phase Transitions. *Journal of Experimental and Theoretical Physics*, *7*, 19–38.

Landau, L. D., & Lifshitz, E. M. (1964). *Statistical physics*. Moscow: Science.

Lehninger, A. L. (1982). *Principles of Biochemistry* (Vol. 1-3). New York: Worth Publisher, Inc.

Lobachevsky, N. (1945). *Geometric research on the theory of parallel lines*. Moscow: Academy of Sciences.

Metzler, D. E. (1980). *Biochemistry. The Chemical Reactions in Living Cells* (Vol. 1-3). New York, San Francisco, London: Academic Press.

Mukamel, D. (1975). Physical relations of n≥. *Physical Review Letters*, *34*, 481–485. doi:10.1103/PhysRevLett.34.481

Mukamel, D., & Krinsky, S. (1976a). Physical relations of n≥. *Physical Review B: Condensed Matter and Materials Physics*, *13*, 5065–5077. doi:10.1103/PhysRevB.13.5065

Mukamel, D., & Krinsky, S. (1976b). Physical relations of n≥. *Physical Review B: Condensed Matter and Materials Physics*, *13*, 5078–5085. doi:10.1103/PhysRevB.13.5078

Mukhopadhyay, N.K., Weatherly, G.C., Lloyd, D.J. & Embury, J.D. (1993). Diffraction studies of icosahedral phases in $Al_{70}Fe_{20}W_{10}$. *Journal of Non-Crystalline Solids, 153-154,* 1193-1197.

Oswald, A. T., MacLeod, C. M., & McCarty, M. (1944). Studies on the Chemical Nature of the Substance Inducing Transformation of Pneumococcal Types: Induction of Transformation by a Desoxyribonucleic Acid Fraction Isolated from Pneumococcus Type III. *The Journal of Experimental Medicine, 79*(2), 137–158. doi:10.1084/jem.79.2.137 PMID:19871359

Patashinskii, A. Z., & Pokrovsky, V. L. (1975). *The fluctuation theory of phase transitions*. Moscow: Science.

Poincare, A. (1895). Analysis situs. *J. de é Ecole Polytechnique, 1*, 1 – 121.

Poincare, A. (1985). *About science*. Moscow: Science.

Ponnuswamy, P., & Gromiha, M. (1994). On the conformational stability of oligonucleotide duplexes and tRNA molecules. *Journal of Theoretical Biology, 169*(4), 419–432. doi:10.1006/jtbi.1994.1163 PMID:7526075

Riemann, B. (1854). *On the Hypotheses Which Lie at the Foundations of Geometry*. Gottingen: Gottingen Observatory.

Shapiro, S. M., Axe, J. D., Shirane, G., & Raccah, P. M. (1974). Neutron scattering study of the structural phase transition in NbO_2. *Solid State Communications, 15*(2), 377–381. doi:10.1016/0038-1098(74)90780-7

Shechtman, D., Blech, I., Gratias, D., & Cahn, J. W. (1984). Metallic phase with longerange orientational order and no translational symmetry. *Physical Review Letters, 53*(20), 1951–1953. doi:10.1103/PhysRevLett.53.1951

Shevchenko, V., Zhizhin, G., & Mackay, A. (2013a). On the structure of the quasicrystals in the high dimension space. In M. V. Diudea (Ed.), *Diamond and Related Nanostructures*. Dordrecht: Springer. doi:10.1007/978-94-007-6371-5_17

Shevchenko, V.Ya., Zhizhin, G.V. & Mackay, A.L. (2013b). On the structure of quasicrystals in a space of higher dimension, *Izvestiya RAS. Chemical series*, 2, 269 - 274.

Voronoi, G. (1908). Recherches sur les parallelloedres primitiss. *Grelle Journal, 134*, 287–1908.

Wilson, R. G. (1971a). Renornalization Group and Critical Phenomena. I. Renormalization Group and the Kadanoff Scaling Picture. *Physical Review B: Condensed Matter and Materials Physics*, *4*(9), 3174–3183. doi:10.1103/PhysRevB.4.3174

Wilson, R. G. (1971b). Renornalization Group and Critical Phenomena. II. Phase – space Cell Analysis of Critical Behavior. *Physical Review B: Condensed Matter and Materials Physics*, *4*(9), 3184–3205. doi:10.1103/PhysRevB.4.3184

Yau, S.-T., & Nadis, S. (2010). *The Shape of Inner Space. String Theory and the Geometry of the Universe's Hidden Dimensions*. New York: Basic Books.

Zhang, X. & Kelton, K.F. (1993). High-order crystal approximant structures alloys $Ti_{54}Zr_{26}Ni_{20}$. *Journal of Non-Crystalline Solids, 153-154*, 114-118.

Zhizhin, G. V. (1972). A qualitative study of one-dimensional steady flows, Leningrad Polytechnic Institute.

Zhizhin, G. V. (2004). *Self-regulating wave of chemical reactions and biological populations.*

Zhizhin, G. V. (2005). *The Dissipative Structures in Chemical, Geological, and Ecological Systems*. S. - Petersburg. *Science*.

Zhizhin, G. V. (2014a). *World – 4D*. St. – Petersburg: Polytechnic Service.

Zhizhin, G. V. (2014b). On the higher dimension in nature. *Biosphere*, *6*(4), 313–318.

Zhizhin, G. V. (2014c). Phase transitions of the second kind with a fluctuation of the geometric structure. *Materials (Basel)*, 10.

Zhizhin, G. V. (2015a). The dimensions of supramolecular compounds. *Biosphere*, *7*(2), 149–154.

Zhizhin, G. V. (2015b). Dimensions of compounds in supramolecular chemistry. *Int. J. Chem. Modeling*, *7*(2), 215–238.

Zhizhin, G. V. (2016a). From multidimensional phase spaces to multidimensionality of space in Nature. *Biosphere*, *8*(3), 258–267.

Zhizhin, G. V. (2016b). The structures, topological and functional dimension of biomolecules. *J Chemoinformatics and Chemical Engineering*, *5*(2), 44–58.

Zhizhin, G. V. (2017). *Chemical Compound Structures and the Higher Dimension of Molecules: emerging Research and Opportunities*. Hershey, PA: IGI Global.

Zhizhin, G. V., & Diudea, M. V. (2016). Space of nanoworld. In M. V. Putz & C. M. Marius (Eds.), *Sustainable nanosystems, development, properties, and applications* (pp. 221–245). New York: IGI Dlobal.

Zhizhin, G. V., Khalaj, Z., & Diudea, M. V. (2016). Geometrical and topology dimensions of the diamond. In A. R. Ashrafi & M. V. Diudea (Eds.), *Distance, symmetry and topology in carbon nanomaterials* (pp. 167–188). New York, London: Springer. doi:10.1007/978-3-319-31584-3_12

KEY TERMS AND DEFINITIONS

Deoxyribonucleic Acid: A biopolymer, the monomer of which is the nucleotide.

Dimension of the Space: The number of independent parameters needed to describe the change in position of an object in space.

Golden Hyper – Rhombohedron: A rhombohedron in a space of higher dimensions, the metric relations between the geometric elements of which are determined by the golden section.

Nucleotide: A phosphoric acid residue attached to sugar deoxyribose, to which one of the four nitrogen bases is attached also.

Phase Portrait: Representation of solutions of a system of differential equations by trajectories in the space of the corresponding phase coordinates.

Polytope: A polyhedron in the space of higher dimension.

Simplex: A convex polytope, any two vertices of which are joined by an edge.

Stereochemistry of Higher Dimension: Representation of the structure of chemical compounds in the space of higher dimensions.

The Order Parameter: A vector characterizing the phase transition of the second kind. Its dimension coincides with the dimension of irreducible representations of the symmetry group of high-symmetry phase.

Zero Surface: The surface at the points of which the derivative of one of the phase coordinates with respect to the independent variable is zero.

Chapter 2
Boundary Complexes and Interior Points of the Polytopes

ABSTRACT

This chapter describes how the structure of a polytope of dimension n consisting of points of the boundary complex including a set of faces from zero to n - 1 and a set of interior points that are not belonging to the boundary complex is considered. The value is equal to the number of elements of the boundary complex, which the given element belongs, having dimension one greater than the given element of the boundary complex is denoted coefficient incidence of the given element. It is proven that the coefficient incidence of an element of dimension i of the boundary complex of an n - cube and n - simplex is equal to the difference of the dimension of the cube or simplex n and the dimension of this element i. The incidence coefficient of elements of n – cross - polytopes is substantially higher than this difference.

INTRODUCTION

A finite family C of convex polytopes will be called a complex under the condition

1. Each face of an element of the family C is an element of the family C;
2. The intersection of any two elements of C is a face of both.

DOI: 10.4018/978-1-5225-6968-8.ch002

Let P is an n – polytope, i.e. a convex polytope of dimension n. We denote by B (P) the boundary complex P, that is, a complex consisting of all faces of P with dimension n - 1 or less (Grunbaum, 1967). The points not belonging to B (P) will be called interior points of P. The set of interior points of the polytope P will be denoted by V. Thus, we have equality $P = B(P) \bigcup V$. Since all the faces of a polytope from vertices (which in chemical compounds are images of atoms, molecules or functional groups) to facets enter the boundary complex, it is of interest to establish relationships between faces of different dimensions in the boundary complex B (P).

For each element b_j of the boundary complex B (P) can to introduce of the notation its dimension $i_{b_j}, 0 \leq i_{b_j} < n$.

There is denoted k_{b_j} the incidence value of the element of the boundary complex b_j to the elements of the boundary complex having dimension $i_{b_j} + 1$. This value is equal to the number of elements of the boundary complex which the given element belongs having dimension one greater than the given element of the boundary complex .

THE STRUCTURE OF THE BOUNDARY COMPLEX OF A N – CUBE

If polytope P is point (vertex), so evidence that n = 0, $i_P = 0, k_P = 0$. If polytope P is segment with vertices b_1, b_2 so the boundary complex B (P) is points b_1, b_2. In this case are $n = 1, i_{b_1} = 0, i_{b_2} = 0, k_{b_1} = 1, k_{b_2} = 1$. Can to introduce the origin of coordinates O in the middle of the segment $[b_1, b_2]$. Thus, if put of length of the segment equal one, so the boundary elements (points) have coordinates $x_{b_1} = -\frac{1}{2}, x_{b_2} = \frac{1}{2}$. The set of interior points in this case is $-\frac{1}{2} < x < \frac{1}{2}$. If n = 2 so polytope P is square (Figure 1). There elements of the boundary complex are vertices b_1, b_2, b_3, b_4 and edges b_1b_2, b_2b_3, b_3b_4, b_4b_1. The dimensions of the vertices equal zero $i_{b_j} = 0, j = 1 \div 4$. The dimensions of the edges equal one $i_{b_1b_2} = i_{b_2b_3} = i_{b_3b_4} = i_{b_4b_1} = 1$. The incidence value of the elements of the boundary complex equal corresponding

$k_{b_j} = 2, j = 1 \div 4, k_{b_1b_2} = k_{b_2b_3} = k_{b_3b_4} = k_{b_4b_1} = 1$. If to introduce the origin of coordinates O in centrum of square and put of length of edges equal one so vertices of square have coordinates (x, y):

$$b_1\left(-\frac{1}{2},\frac{1}{2}\right), b_2\left(\frac{1}{2},\frac{1}{2}\right), b_3\left(\frac{1}{2},-\frac{1}{2}\right), b_4\left(-\frac{1}{2},-\frac{1}{2}\right).$$

The edges are defined by the relations

$$b_1b_2 = \left\{y = \frac{1}{2}; -\frac{1}{2} \le x \le \frac{1}{2}\right\}; b_2b_3 = \left\{x = \frac{1}{2}; -\frac{1}{2} \le y \le \frac{1}{2}\right\};$$

$$b_3b_4 = \left\{y = -\frac{1}{2}; -\frac{1}{2} \le x \le \frac{1}{2}\right\}; b_4b_1 = \left\{x = -\frac{1}{2}; -\frac{1}{2} \le y \le \frac{1}{2}\right\}.$$

The set of interior points V is defined by the inequalities $-\frac{1}{2} < x < \frac{1}{2}, -\frac{1}{2} < y < \frac{1}{2}$.

Figure 1. The square

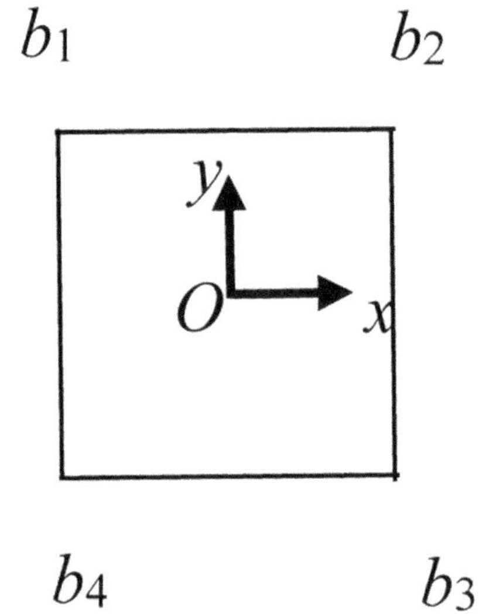

Figure 2. The cube

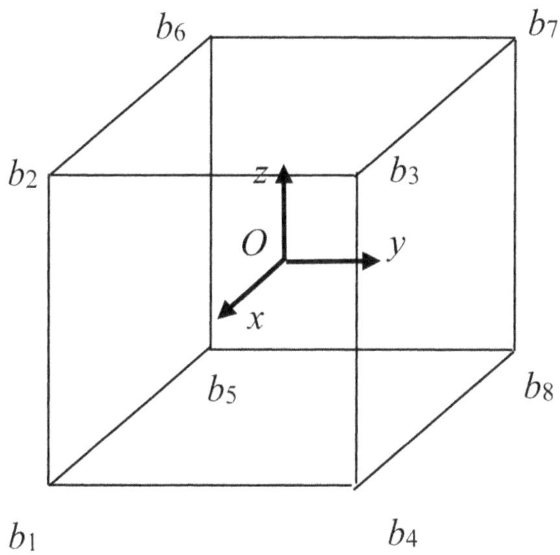

At $n = 3$ in boundary complex of polytope P the 3 – cube (Figure 2) entre vertices $b_j (j = 1 \div 8)$; edges

$$e_1 = b_1 b_2, e_2 = b_2 b_3, e_3 = b_3 b_4, e_4 = b_4 b_1, e_5 = b_5 b_6, e_6 = b_6 b_7, e_7 = b_7 b_8,$$

$$e_8 = b_5 b_8, e_9 = b_1 b_5, e_{10} = b_4 b_8, e_{11} = b_2 b_6, e_{12} = b_3 b_7;$$

flat faces (squares)

$$s_1 = b_1 b_2 b_3 b_4, s_2 = b_5 b_6 b_7 b_8, s_3 = b_5 b_1 b_4 b_8, s_4 = b_1 b_2 b_5 b_6, s_5 = b_2 b_6 b_3 b_7, s_6 = b_8 b_4 b_3 b_7.$$

It should be noted that the edges are images of valence bonds in chemical compounds, but in addition to they create a skeleton of a convex figure. In this case, part of the edges carries only this geometric function. It is evident, that

$i_{b_j} = 0, j = 1 \div 8; i_{e_j} = 1, j = 1 \div 12; k_{b_j} = 3, j = 1 \div 8; k_{e_j} = 2, j = 1 \div 12; k_{s_j} = 1, j = 1 \div 6.$

If to introduce the origin of coordinates O in centrum of cube (Figure 2) and put of length of edges equal one so vertices of cube have coordinates (x, y, z):

$$b_1\left(\frac{1}{2},-\frac{1}{2},-\frac{1}{2}\right), b_2\left(\frac{1}{2},-\frac{1}{2},\frac{1}{2}\right), b_3\left(\frac{1}{2},\frac{1}{2},\frac{1}{2}\right), b_4\left(\frac{1}{2},\frac{1}{2},-\frac{1}{2}\right),$$

$$b_5\left(-\frac{1}{2},-\frac{1}{2},-\frac{1}{2}\right), b_6\left(-\frac{1}{2},-\frac{1}{2},\frac{1}{2}\right), b_7\left(-\frac{1}{2},\frac{1}{2},\frac{1}{2}\right), b_8\left(-\frac{1}{2},\frac{1}{2},-\frac{1}{2}\right).$$

The total number of vertices is equal to 2^3, since at each place of the chain of coordinates values of length 3 there can be two values $\pm\frac{1}{2}$. The edges are determined by the conditions for the equality $\pm\frac{1}{2}$ if any two variables and the change from - $\frac{1}{2}$ to + $\frac{1}{2}$ the third variable. Flat faces are defined by the equality $\pm\frac{1}{2}$ of any one of the variables and changing from - $\frac{1}{2}$ to +$\frac{1}{2}$ two other variables. The set of interior points V is determined by a set of inequalities $-\frac{1}{2} < x < \frac{1}{2}, -\frac{1}{2} < y < \frac{1}{2}, -\frac{1}{2} < z < \frac{1}{2}$.

At $n = 4$ in boundary complex of polytope P the 4 – cube (Figure 3) entre vertices, edges, quadrat faces and the 3 – cubes (Stringham, 1880).

Figure 3. The 4 – cube

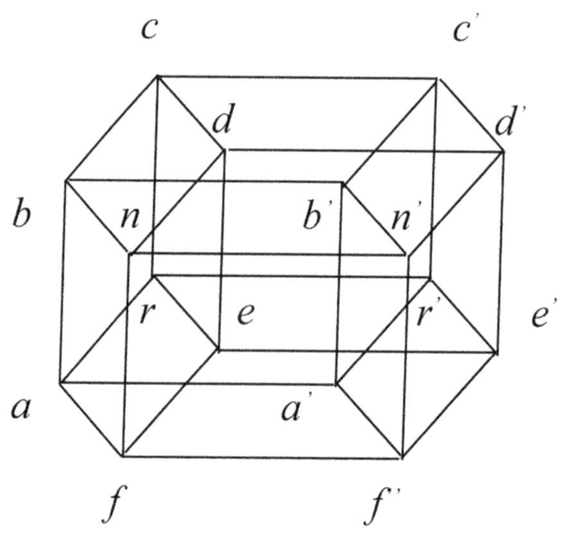

The number of vertices is $2^4 = 16$. If the origin of the coordinates is placed in the center of the cube, and the length of the edge is taken as one, then the coordinates of the vertices are determined by different locations of $\pm\frac{1}{2}$ in the four places of the chain (*x, y, z, t*). The edges are determined by the conditions for the equality $\pm\frac{1}{2}$ of any 3 variables and the change from $-\frac{1}{2}$ to $+\frac{1}{2}$ the fourth variable. The number of the edges equal $2^3 4 = 32$. Flat faces it are defined by the equality $\pm\frac{1}{2}$ of any two of the variables and changing from $-\frac{1}{2}$ to $+\frac{1}{2}$ two other variables. The number of the flat faces equal to $2^2 \cdot 6 = 24$. The three - dimensional faces are defined by the equality $\pm\frac{1}{2}$ of one of the four variables and the change from $-\frac{1}{2}$ to $+\frac{1}{2}$ the remaining variables. The number of the three - dimensional faces equal $2 \cdot 4 = 8$. For the following is more convenient to denote the elements of a boundary complex with an indication of its dimension $b(i)$. It follows from the Figure 3 that the incidence value of the element with dimension equal zero

$$k_{b(i=0)}(4) = 4, k_{b(i=1)}(4) = 3, k_{b(i=2)}(4) = 2, k_{b(i=3)}(4) = 1.$$

On the base of the consideration of the relations between elements of different dimensions in *n* - cubes ($n = 0 \div 4$), can to assert the following.

For arbitrary *n*, the number of vertices in the *n* - cube, i.e. elements of dimension zero, is equal to $f_0 = 2^n$. The number of edges, i.e. elements one dimension, it defined by equality $\pm\frac{1}{2}$ any of *n*-1 variables and changes from $-\frac{1}{2}$ to $+\frac{1}{2}$ remaining one variable. This number is $f_1 = 2^{n-1}C_n^1, C_n^1$ -the number combination from *n* elements on one. The number of elements with dimension two is defined by the equality $\pm\frac{1}{2}$ of any $n - 2$ variables and changes from $-\frac{1}{2}$ to $+\frac{1}{2}$ remaining two variables, i.e. $f_2 = 2^{n-2}C_n^2$ and so on.

So we get a general formula for calculating the number of elements of dimension i in the n – cube

$$f_i = 2^{n-i} C_n^i, i = 0, 1, 2, \ldots, n. \qquad (1)$$

This formula has been obtained previously (Zhizhin, 2013, 2014a) by other means, recurrently examining relationship between elements of different dimensions by increasing the dimension of the polytope unit. It can be shown (Zhizhin, 2014a) that formula (1) satisfies the Euler - Poincaré equation (Poincaré, 1895)

$$\sum_{i=0}^{n-1} (-1)^i f_i(P) = 1 + (-1)^{n-1}, \qquad (2)$$

the fulfillment of which is necessary for the existence of a closed convex polytope of dimension n.

Theorem 1

The incidence of an element of dimension i of the boundary complex of an n - cube is equal to the difference of the dimension of the cube n and the dimension of this element, i.e.

$$k_{b(i)} = n - i. \qquad (3)$$

Proof

The proof of this theorem is the analysis of the incidence coefficient of the elements of the boundary complex of n - polytopes for various values of n, in particular, the analysis already performed for $n = 0, 1, 2, 3, 4$. In each of these polytopes, the incidence coefficient of the elements of the boundary complex is equal to the difference in the dimension of the polytope and the dimension of the element of the boundary complex itself. This equality holds for other n. For example, for $n = 5$, the incidence coefficient values for elements of a smaller - dimensional boundary complex calculated by formula (3) equal

Figure 4. The 5 - cube

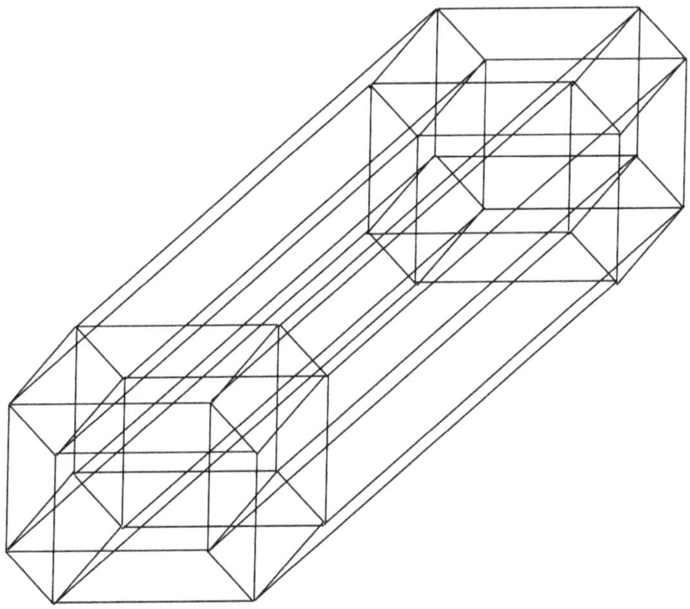

$$k_{b(i=0)}(5) = 5, k_{b(i=1)}(5) = 4, k_{b(i=2)}(5) = 3, k_{b(i=3)}(5) = 2, k_{b(i=4)}(5) = 1$$

In Figure 4, a 5 - polytope is constructed (Zhizhin, 2014a). This figure confirms the incidence values calculated by formula (3). It can be seen that formula (3) is also satisfied for large values of *n*.

Theorem 1 is proved.

In particular, it follows from Theorem 1 that as the dimension of the *n* - cube grows, the number of the 3 - cubes having a common plane face increases proportionally linearly. In a 4 - cube, a planar face is common for two the 3 - cubes, in a 5 - cube a flat face is common for three the 3 - cubes, and so on.

In any *n* - cube, there are many interior points that do not belong to any facet. This set is determined by a set of inequalities (if the origin is placed in the center of the *n* - cube, and the length of the edge is taken as one)

$$-\frac{1}{2} < x_1 < \frac{1}{2}, -\frac{1}{2} < x_2 < \frac{1}{2}, \ldots, -\frac{1}{2} < x_n < \frac{1}{2}.$$

THE STRUCTURE OF THE BOUNDARY COMPLEX OF A REGULAR N – SIMPLEX

At $n = 2$ the polytope P regular 2 - simplex is a regular triangle (Figure 5). In the boundary complex of the 2 – simplex entre the vertices $b_j, j = 1,2,3;$ edges $e_j, j = 1,2,3; e_1 = b_1b_2; e_2 = b_2b_3; e_3 = b_3b_1.$

The dimension of the vertices equal zero $i_{b_j} = 0, j = 1 \div 3.$ The dimension of the edges equal one $i_{e_j} = 1, j = 1 \div 3.$ The incidence value of the elements of the boundary complex equal corresponding $k_{b_j} = 2, k_{e_j} = 1, j = 1 \div 3.$ If to introduce the origin of coordinates O in centrum of triangle and put of length of edges equal one so vertices of triangle have coordinates (x, y): $b_1\left(0, \frac{1}{\sqrt{3}}\right), b_2\left(\frac{1}{2}, -\frac{1}{2\sqrt{3}}\right), b_3\left(-\frac{1}{2}, -\frac{1}{2\sqrt{3}}\right).$

The equations of the edges in this system of coordinates have view $e_1 : y = -\sqrt{3}x + \frac{1}{\sqrt{3}}; e_2 : y = -\frac{1}{2\sqrt{3}}; e_3 : y = \sqrt{3}x + \frac{1}{\sqrt{3}}.$ Consequently, the domain of interior points is determined by the system of inequalities

Figure 5. The triangle

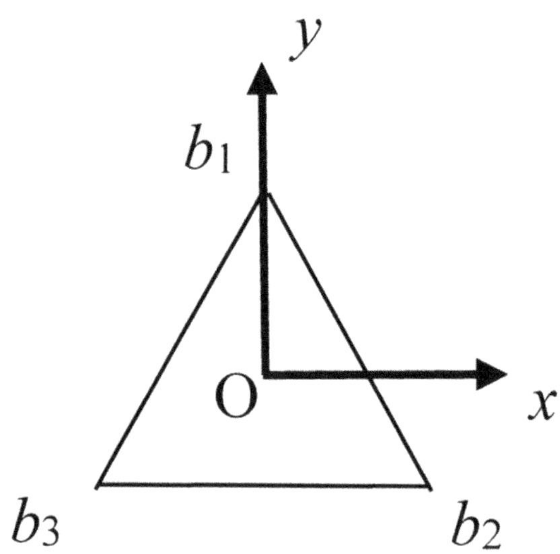

$$y < -\sqrt{3}x + \frac{1}{\sqrt{3}}, 0 < x < \frac{1}{2};$$

$$y < \sqrt{3}x + \frac{1}{\sqrt{3}}, -\frac{1}{2} < x < 0;$$

$$y > -\frac{1}{2\sqrt{3}}, -\frac{1}{2} < x < 0.$$

It is evident, that $i_{b_j} = 0, i_{e_j} = 1, k_{b_j} = 2, k_{e_j} = 1, j = 1 \div 3.$

At $n = 3$ the polytope P regular 3 - simplex is a regular tetrahedron (Figure 6).

In the boundary complex of the 3 – simplex entre the vertices $b_j, j = 1 \div 4;$ theedges $e_j, j = 1 \div 6, e_1 = b_1b_2, e_2 = b_2b_3, e_4 = b_4b_1, e_5 = b_4b_2, e_6 = b_3b_4;$ theregular triangles $s_j, j = 1 \div 4, s_1 = b_1b_2b_3, s_2 = b_1b_2b_4, s_3 = b_2b_3b_4, s_4 = b_1b_3b_4.$

The dimension of the vertices equal zero $i_{b_j} = 0, j = 1 \div 4.$

Figure 6. The tetrahedron

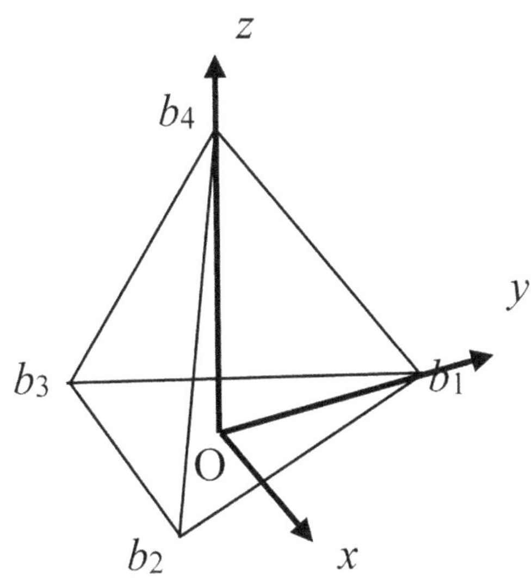

The dimension of the edges equal one $i_{e_j} = 1, j = 1 \div 6$. The dimension of the regular triangles equal two $i_{s_j} = 2, j = 1 \div 4$. The incidence value of the elements of the boundary complex equal corresponding $k_{b_j} = 3, j = 1 \div 4; k_{e_j} = 2, j = 1 \div 6; k_{s_j} = 1, j = 1 \div 4$. If to introduce the origin of coordinates *O* in centrum of triangle s_1 and put of length of edges equal one so vertices of tetrahedron have coordinates (*x*, *y*, *z*)

$$b_1\left(0, \frac{1}{\sqrt{3}}, 0\right), b_2\left(\frac{1}{2}, -\frac{1}{2\sqrt{3}}, 0\right), b_3\left(-\frac{1}{2}, -\frac{1}{2\sqrt{3}}, 0\right), b_4\left(0, 0, \sqrt{\frac{2}{3}}\right).$$

The equations of flats S_j in which lie the triangles s_j, $j = 1 \div 4$, have view

$$S_1 : z = 0,$$

$$S_2 : z = -\sqrt{6}x - \sqrt{2}y + \sqrt{\frac{2}{3}},$$

$$S_3 : z = 2\sqrt{2}y + \sqrt{\frac{2}{3}},$$

$$S_4 : z = 2\sqrt{\frac{2}{3}}x - \sqrt{2}y + \sqrt{\frac{2}{3}}.$$

In connection with this, the set of interior points of a tetrahedron is determined by a set of inequalities

$$z > 0;$$

$$z < 2\sqrt{\frac{2}{3}}x - \sqrt{2}y + \sqrt{\frac{2}{3}}, -\frac{1}{2} < x < 0, -\frac{1}{2\sqrt{3}} < y < \frac{1}{\sqrt{3}};$$

$$z < 2\sqrt{2}y + \sqrt{\frac{2}{3}}, -\frac{1}{2} < x < \frac{1}{2}; -\frac{1}{2\sqrt{3}} < y < 0;$$

Figure 7. The 4 - simplex

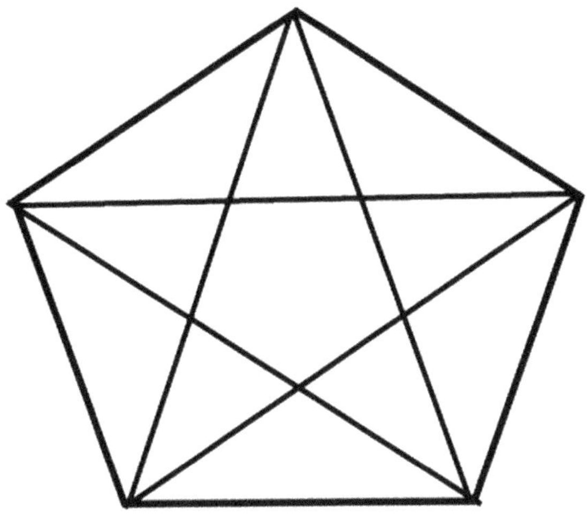

$$z < -\sqrt{6}x - \sqrt{2}y + \sqrt{\frac{2}{3}}, 0 < x < \frac{1}{2}, -\frac{1}{2\sqrt{3}} < y < \frac{1}{\sqrt{3}}.$$

At $n = 4$ in boundary complex of polytope P the 4 – simplex (Figure 7) entre five vertices $b(i = 0)$, ten edges $b(i = 1)$, ten flat triangular faces $b(i = 2)$ and five tetrahedrons $b(i = 3)$ (Stringham, 1880).

The incidence coefficient values for elements of a boundary complex of the 4 – simplex equal

$$k_{b(i=0)}(4) = 4, k_{b(i=1)}(4) = 3, k_{b(i=2)}(4) = 2, k_{b(i=3)}(4) = 1.$$

Putting the origin of the coordinates (x, y, z, t) in the center of the 4 - simplex, in principle, can to write the coordinates of the vertices of the simplex and define the region of interior points using linear programming methods in a multidimensional space (Vasil'ev, Ivanitsky, 2008). In view of the awkwardness of these calculations in this and similar cases, shall not give.

At $n = 5$ in boundary complex of polytope P 5 – simplex (Figure 8) entre six vertices $b(i = 0)$, 15 edges $b(i = 1)$, 20 flat triangular faces $b(i = 2)$, 15 tetrahedrons $b(i = 3)$ and six 4 – simples (Zhizhin, 2013, 2014a).

A general formula for calculating the number of elements of dimension i in n – simplex is (Zhizhin, 2014a)

Figure 8. The 5 – simplex

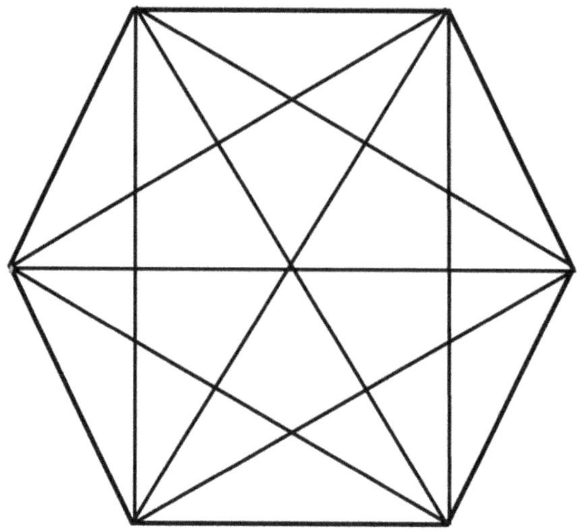

$$f_i(n) = C_{n+1}^{i+1}, i = 0, 1, 2, ..., n. \quad (4)$$

This formula has been obtained previously (Zhizhin, 2013, 2014a) by other means. The incidence coefficient values for elements of a boundary complex 5 –simplex equal

$$k_{b(i=0)}(5) = 5, k_{b(i=1)}(5) = 4, k_{b(i=2)}(5) = 3, k_{b(i=1)}(5) = 2, k_{b(i=0)}(5) = 1.$$

Thus, just as in *n* - cubes in *n* - simplexes, the coefficients of incidence of the elements of the boundary complex increase linearly with the increase in the dimension of the polytope. This allows us to formulate a more general statement:

Theorem 2

The incidence of an element of dimension *i* of the boundary complex of an *n* - cube and *n* – simplex is equal to the difference of the dimension of the polytope and the dimension of this element, i.e.

$$k_{b(i)}(n) = n - i. \quad (5)$$

Remark: It is surprising that the same difference is equal to the incidence coefficient of the dimension element and in the n - simplex with respect to the facets of this polytope (see Chapter 5)

$k_{b(i),b(n-1)}(n) = n - i.$ (5)[a]

Really, in a 4 – simplex are

$k_{b(0),b(3)} = 4; k_{b(1),b(3)} = 3; k_{b(2)b(3)} = 2;$

In a 5 – simplex are

$k_{b(0),b(4)} = 5; k_{b(1),b(4)} = 4; k_{b(2),b(4)} = 3; k_{b(3),b(4)} = 2,$

and go on.

THE STRUCTURE OF THE BOUNDARY COMPLEX OF A REGULAR N – CROSS -POLYTOPE

The structure of the boundary complex of a regular the n – cross – polytope is essentially differed from the structures of the n - cube and of the n – simplex. The number of the elements of different dimension of the n – cross – polytope can to count on equation (Zhizhin, 2013, 2014a)

$$f_i(n) = 2^{1+i} C_n^{n-1-i}. \qquad (6)$$

According to equation (6) for the n – cross - polytope of the smallest possible dimension 4, we have $f_0(4) = 8, f_1(4) = 24, f_2(4) = 32, f_3(4) = 16$. Thus, the 4 – cross – polytope has 8 vertices, 24 edges, 32 flat faces (triangles), 16 tetrahedrons. On Figure 9 it is shown the 4 – cross – polytope in edited of Coxeter (Coxeter, 1963). The peculiarity of the n – cross - polytope is that in its projection onto the two - dimensional plane, the vertices opposite to the center of the polytope are not connected by an edge.

Figure 9. The 4 – cross - polytope

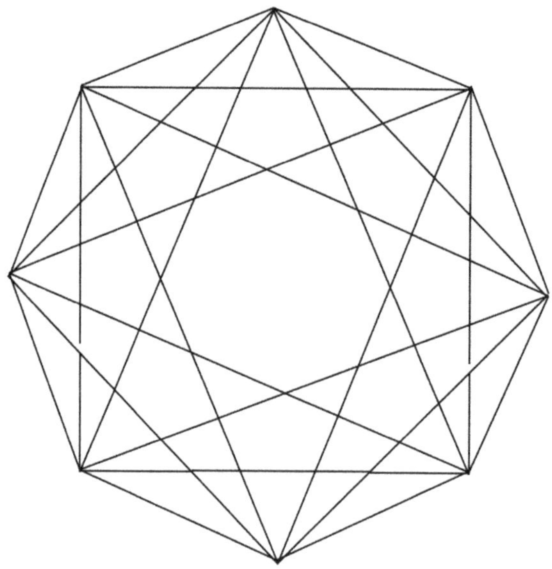

It follows from Figure 9 that the incidence coefficients of the elements of the boundary complex the 4 – cross - polytope have the values

$$k_{b(i=0)}(4) = 6, k_{b(i=1)}(4) = 4, k_{b(i=2)}(4) = 2, k_{b(i=3)}(4) = 1$$

Comparing incidence coefficients of the elements of the n - cube, n - simplex and n – cross -polytope boundary complex for the same value of $n = 4$, can to see that the incidence coefficients of the n – cross - polytope of elements of dimension 0 and 1 are higher than the corresponding incidence values of the 4 - simplex and of the 4 - cube. This difference in the incidence values increases with increasing n. In addition, with increasing n, the difference in the values of the incidence coefficients of the n - cubes, n - simplexes, and n – cross - polytopes extends to elements of an increasingly dimension. For example, according to equation (6) for the 5 – cross - polytope are

$$f_0(5) = 10, f_1(5) = 40, f_2(5) = 80, f_3(5) = 80, f_4(5) = 32.$$

Thus, the 5 – cross – polytope has 10 vertices, 40 edges, 80 flat faces (triangles), 80 tetrahedrons and 32 polytopes with dimension 4. On Figure

Figure 10. The 5 – cross - polytope

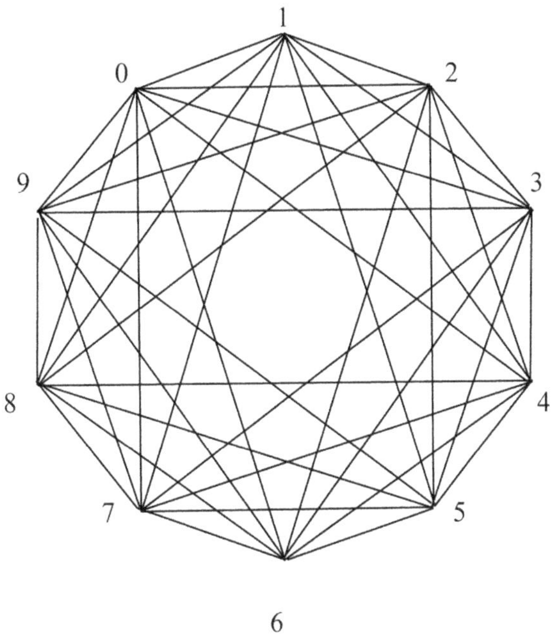

10 it is shown 5 – cross – polytope corresponding this values the elements with different dimension.

It follows from Figure 10 that the incidence coefficients of the elements of the boundary complex 5 – cross - polytope have the values

$$k_{b(i=0)}(5) = 8, k_{b(i=1)}(5) = 6, k_{b(i=2)}(5) = 4, k_{b(i=3)}(5) = 2, k_{b(i=4)}(5) = 1$$

According to equation (6) for the 6 – cross - polytope are

$$f_0(6) = 12, f_1(6) = 60, f_2(6) = 160, f_3(6) = 240, f_4(6) = 192, f_5(6) = 64.$$

Thus, the 6 – cross – polytope has 12 vertices, 60 edges, 160 flat faces (triangles), 240 tetrahedrons, 192 polytopes with dimension 4 and 64 polytopes with dimension 5. On Figure 11 it is shown the 6 – cross – polytope corresponding this values the elements with different dimension.

It follows from Figure 11 that the incidence coefficients of the elements of the boundary complex 6 – cross - polytope have the values

Figure 11. The 6 – cross - polytope

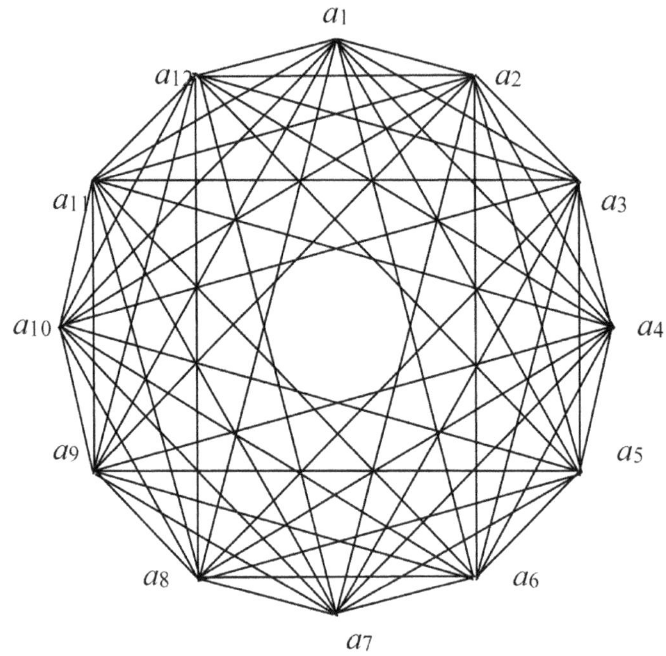

$$k_{b(i=0)}(6) = 10, k_{b(i=1)}(6) = 8, k_{b(i=2)}(6) = 6, k_{b(i=3)}(6) = 4, k_{b(i=4)}(6) = 2, k_{b(i=5)}(6) = 1$$

From the analysis carried out follows the statement

Theorem 3

The incidence coefficients of the elements of the boundary complex of n – cross -polytopes of dimension i if $i < n - 1$ are defined by the equality $k_{b(i)}(n) = 2(n-1-i)$. If $i = n - 1$, so

$$k_{b(n-1)}(n) = 1. \tag{7}$$

Significantly larger values of the incidence coefficients of the elements of the boundary complex of n – cross - polytopes as compared to the n - cubes and n - simplexes should be taken into account in the study of structures involving the n – cross - polytopes (Chapter 6).

It should be noted that, whatever the size of the polytope, if it is surrounded by three-dimensional space, then the outer boundary of the polytope will be a three - dimensional body with two - dimensional sides. The existence of a polytope of higher dimension inside a three -dimensional space does not contradict Riemann's geometry (Riemann, 1854), which assumes the boundedness of a space of higher dimension (Zhizhin, 2014b).

BOUNDURY COMPLEXES OF THE CLUSTERS

Cluster of Mackay

The Mackay cluster is two icosahedra with a common center of different sizes. The smaller icosahedron is embedded in the larger one, and each vertex of the larger icosahedron is on the same line passing through the common center, with the vertex of the smaller icosahedron (Lord, Mackay & Ranganathan, 2006). At the vertices of each icosahedron are atoms, in addition, in the middle of each side of the triangular face of a larger icosahedron, too, are atoms. To determine the cluster dimension, these atoms, due to their linear arrangement, have no effect. The other vertices are important. They are 24, on 12 from each

Figure 12. Cluster of Mackay

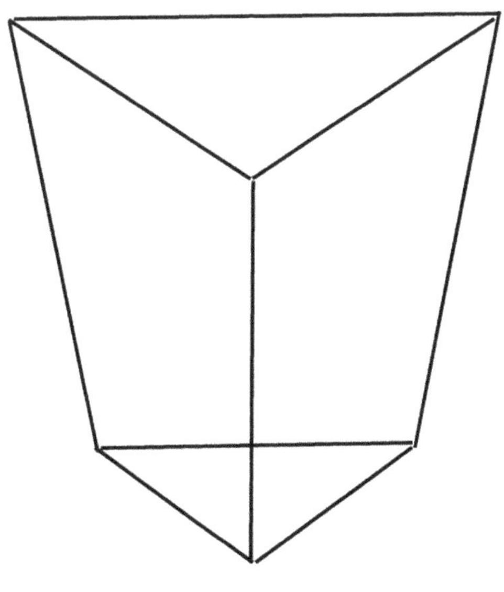

icosahedron. One join the edges of the vertices of both icosahedron lying on the line passing through the center of the icosahedrons. Then the space between the icosahedrons will be filled with triangular prisms, the bases of which are the triangular faces of the larger and smaller icosahedron (Figure 12).

These prisms are adjacent to each other along entire flat lateral faces. The number of these prisms is equal to the number of triangular faces of the icosahedron, i.e. 20.

In this case the number of vertices equal $f_0 = 24$.

The number of edges equal sum of the number edges at two icosahedrons and the number edges consequent of vertices both icosahedrons: $f_1 = 60 + 12 = 72$.

The flat faces is sum of the number flat faces of icosahedrons and the number of flat faces between icosahedrons: $f_2 = 40 + 30 = 70$. The number of tree – dimension faces is sum of the number of the prisms and the number of icosahedrons: $f_3 = 20 + 2 = 22$.

Substituting the calculated values of the numbers of faces of different dimensions into Euler – Poincare equation (2), can to obtain

$$24 - 72 + 70 - 22 = 0,$$

Thus, the Euler - Poincare equation is satisfied at a value of n equal to 4. Hence, the cluster of Mackay has dimension 4.

The cluster of γ - latten

For construction of γ - latten (see (Lord, Mackay & Ranganathan, 2006)) it need to faces of tetrahedron (Figure 13) joined yet four tetrahedrons (Figure 14).

The space between the tetrahedrons is suggested to be filled by three double tetrahedrons, i.e. six tetrahedrons). However, there is an easier way to get a closed figure. It suffices to inscribe one tetrahedron 5678 (red edges) into the Figure 14. The result of this procedure is the polytope in Figure 15.

It has 8 vertices, each of them has six edges. So, polytope contains 16 tetrahedrons. This is a 4 – cross - polytope. Thus, the dimension of the simplest cell of the cluster of γ - latten is 4. To this polytope there is possible to attach more tetrahedrons, then polytopes of even greater dimension are formed.

Often, one or more atoms are located in the center of the clusters. It is of interest to consider the features of the structure of polytopes of higher dimension, taking into account such an arrangement of atoms.

Figure 13. Initial tetrahedron

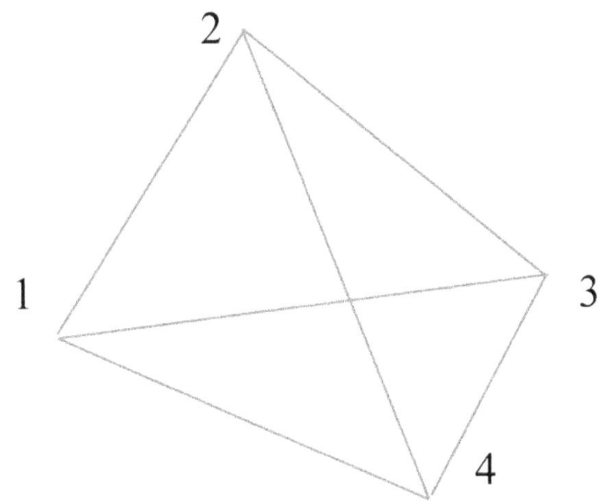

Figure 14. Initial tetrahedron with joined tetrahedrons

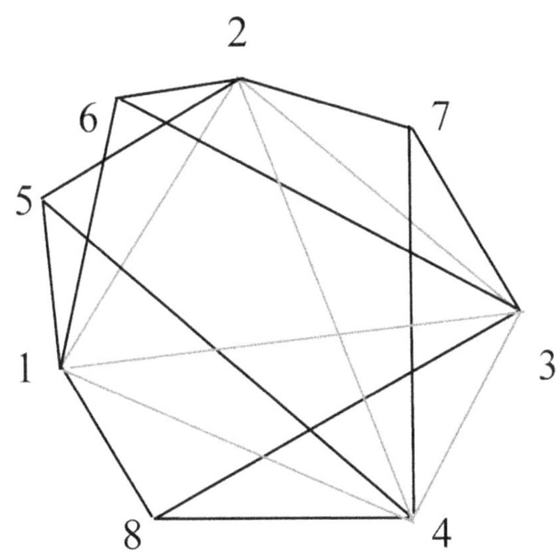

Figure 15. The cluster of γ - latten

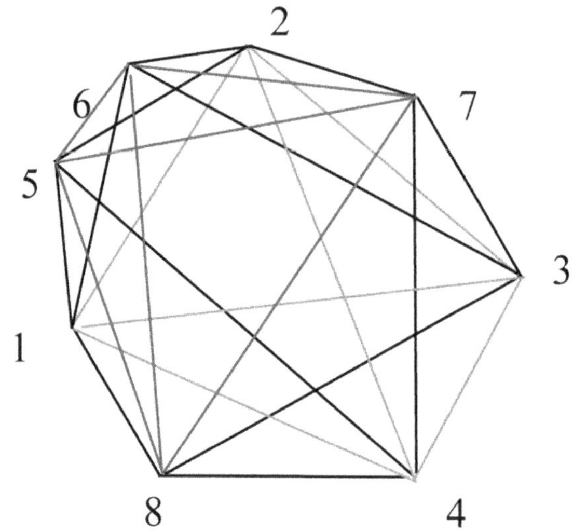

Figure 16. Cube with center

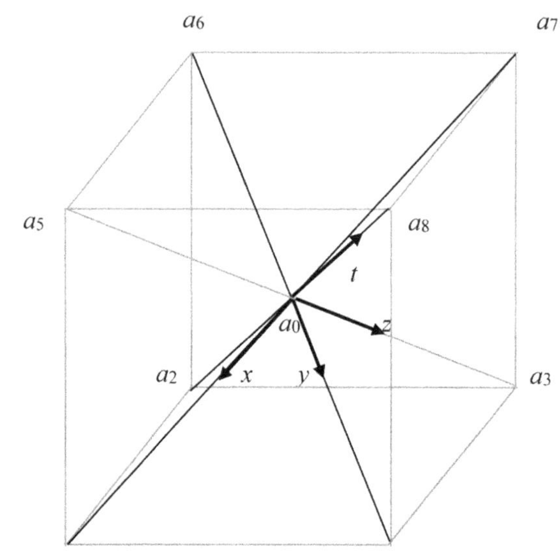

The Structure of Boundary Complex of the Cube With Center

Consider a cube with center at point a_0 (Figure 16).

It is denoted the vertices of the cube $a_j (j = 1 \div 8)$ and draw through the center of the diagonals of the cube. Let's enumerate the geometric elements of different dimensions of this construction: 9 vertices $(f_0 = 9)$; 20 edges ($f_1 = 20$):

$$e_1 = a_1a_2, e_2 = a_3a_2, e_3 = a_3a_4, e_4 = a_1a_4, e_5 = a_5a_6, e_6 = a_6a_7, e_7 = a_7a_8, e_8 = a_5a_8, e_9 = a_1a_5,$$
$$e_{10} = a_4a_8, e_{11} = a_6a_2, e_{12} = a_3a_7, e_{13} = a_0a_1, e_{14} = a_0a_2, e_{15} = a_0a_3, e_{16} = a_0a_4, e_{17} = a_0a_5,$$
$$e_{18} = a_0a_6, e_{19} = a_0a_7, e_{20} = a_0a_8;$$

18 two – dimensional faces ($f_2 = 18$): 6 square faces –

$$s_1 = a_1a_2a_3a_4, s_2 = a_5a_6a_7a_8, s_3 = a_1a_5a_8a_4, s_4 = a_6a_2a_3a_7, s_5 = a_1a_2a_5a_6, s_6 = a_7a_8a_3a_4;$$

12 triangular faces –

$$s_7 = a_0a_1a_4, s_8 = a_0a_3a_4, s_9 = a_0a_2a_3, s_{10} = a_0a_1a_2, s_{11} = a_0a_5a_8, s_{12} = a_0a_5a_6, s_{13} = a_0a_6a_7,$$
$$s_{14} = a_0a_8a_7, s_{15} = a_0a_6a_2, s_{16} = a_0a_1a_5, s_{17} = a_0a_3a_7, s_{18} = a_0a_8a_4;$$

7 three - dimensional faces ($f_3 = 7$):
1 cube, 6 pyramids –

$$v_2 = a_0a_1a_2a_3a_4, v_3 = a_0a_5a_6a_7a_8, v_4 = a_0a_1a_2a_5a_6, v_5 = a_0a_7a_8a_3a_4, v_6 = a_0a_1a_5a_8a_4, v_7 = a_0a_6a_2a_3a_7$$

Substituting the obtained quantities of elements of different dimensions into the Euler - Poincare equation (2), can to obtain

9 - 20 + 18 - 7 = 0

This means that equation (2) is satisfied for $n = 4$. Consequently, the cube with the center has dimension 4 (Zhizhin & Diudea, 2017).

The incidence of the elements of the boundary complex in this case is determined by the equalities

$$k_{a_0}(4)=8; k_{a_j}(4)=4, j=1\div 8; k_{e_j}(4)=3, j=1\div 20; k_{s_j}(4)=2, j=1\div 18; k_{v_j}(4)=1, j=1\div 7$$

Since the cube with the center has dimension 4, it is necessary to place the origin of coordinates of the four - dimensional system in it. It is convenient to take the center of the cube as the origin of coordinates by sending coordinates along the diagonals of the cube (Figure 16).

Let's take a guess from the center of the cube to any of its vertices for a unit. Then all the vertices of polytope are determined by the values of the coordinates (*x, y, z, t*)

$$a_0\left(0,0,0,0\right), a_1\left(1,0,0,0\right), a_2\left(0,0,0,-1\right), a_3\left(0,0,1,0\right), a_4\left(0,1,0,0\right), a_5\left(0,0,-1,0\right),$$
$$a_6\left(0,-1,0,0\right), a_7\left(-1,0,0,0\right), a_8\left(0,0,0,1\right).$$

Each of the facets of a boundary complex is determined by the corresponding system of inequalities

$$v_1: -1\le x\le 1, -1\le y\le 1, -1\le z\le 1, -1\le t\le 1; v_2: 0\le x\le 1, 0\le y\le 1, 0\le z\le 1, -1\le t\le 0;$$
$$v_3: -1\le x\le 0, -1\le y\le 0, -1\le z\le 0, 0\le t\le 1; v_4: 0\le x\le 1, -1\le y\le 0, -1\le z\le 0, -1\le t\le 0;$$
$$v_5: -1\le x\le 0, 0\le y\le 1, 0\le z\le 1, 0\le t\le 1; v_6: 0\le x\le 1, 0\le y\le 1, -1\le z\le 0, 0\le t\le 1;$$
$$v_7: -1\le x\le 0, -1\le y\le 0, 0\le z\le 1, 0\le t\le 1.$$

Thus, the entire polytope it is occupied by its facets, i.e. its boundary complex, the polytope does not have interior points. If one recall that the boundary complex is used as a synonym for the surface of the figure, then we came to an amazing result: the surface of the cube with the center completely occupies the space of this polytope.

The Structure of Boundary Complex of the Octahedron With Center

The octahedron with the center has 7 vertices (Figure 17) $a_0\div a_6, (f_0=7)$; 18 edges $\left(f_1=18\right)$

Figure 17. The octahedron with center

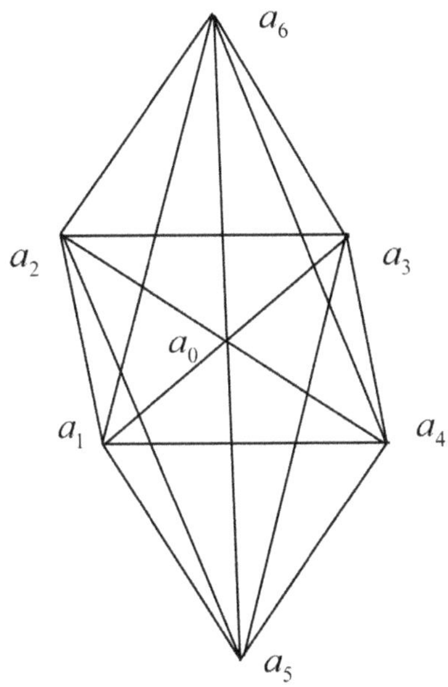

$e_1 = a_0a_3, e_2 = a_0a_4, e_3 = a_0a_1, e_4 = a_0a_2, e_5 = a_0a_6, e_6 = a_0a_5, e_7 = a_1a_6, e_8 = a_2a_6, e_9 = a_6a_3,$;
$e_{10} = a_4a_6, e_{11} = a_1a_5, e_{12} = a_2a_5, e_{13} = a_4a_5, e_{14} = a_5a_3, e_{15} = a_1a_2, e_{16} = a_2a_3, e_{17} = a_4a_3, e_{18} = a_1a_4,$

20 flat two – dimensional trigonal faces $\left(f_2 = 20\right)$-

$s_1 = a_1a_2a_6, s_2 = a_3a_2a_6, s_3 = a_3a_4a_6, s_4 = a_1a_4a_6, s_5 = a_1a_2a_5, s_6 = a_3a_2a_5, s_7 = a_3a_4a_5,$
$s_8 = a_1a_4a_5, s_9 = a_1a_0a_6, s_{10} = a_0a_2a_6, s_{11} = a_0a_3a_6, s_{12} = a_0a_4a_6, s_{13} = a_1a_0a_5, s_{14} = a_0a_4a_5,$
$s_{15} = a_0a_3a_5, s_{16} = a_0a_2a_5, s_{17} = a_1a_0a_4, s_{18} = a_1a_2a_0, s_{19} = a_0a_2a_3, s_{20} = a_0a_3a_4;$

9 three – dimensional faces $\left(f_3 = 9\right)$: 1 the octahedron, 8 the pyramids

$v_2 = a_1a_0a_4a_5, v_3 = a_3a_0a_4a_5, v_4 = a_3a_0a_2a_5, v_5 = a_1a_0a_2a_5, v_6 = a_1a_0a_4a_6, v_7 = a_3a_0a_4a_6,$
$v_8 = a_3a_0a_2a_6, v_9 = a_1a_0a_2a_6.$

Substituting the obtained quantities of elements of different dimensions into the Euler - Poincare equation (2), can to obtain

7 - 18 + 20 - 9 = 0.

This means that equation (2) is satisfied for $n = 4$. Consequently, the octahedron with the center has dimension 4.

The incidence of the elements of the boundary complex in this case is determined by the equalities

$$k_{a_0}(4) = 6; k_{a_j}(4) = 5, j = 1 \div 6; k_{e_j}(4) = 4, j = 1 \div 6; k_{e_j}(4) = 3, j = 7 \div 18;$$
$$k_{s_j}(4) = 2, j = 1 \div 20; k_{v_j}(4) = 1, j = 1 \div 9.$$

Now can to see, without introducing the coordinate system (Figure 17), that an octahedron with a center does not have interior points, since the entire neighborhood of the center is filled with facets.

The structure of boundary complex of the tetrahedron with center

Figure 18. The tetrahedron with center

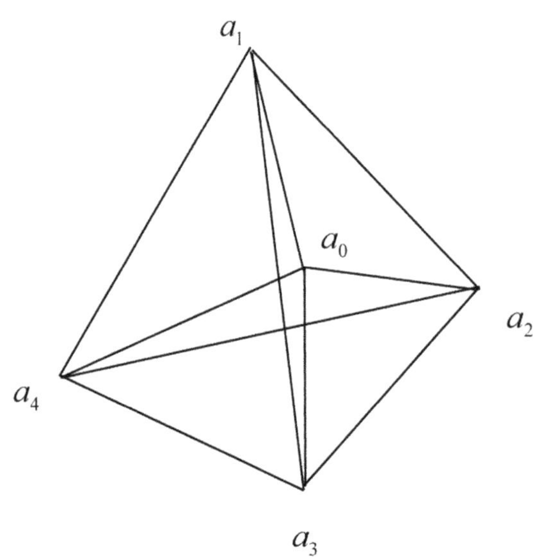

The tetrahedron with the center has 5 vertices (Figure 18) $a_0 \div a_5, (f_0 = 5)$; 10 edges $(f_1 = 10)$, 10 two – dimensional trigonal faces $(f_2 = 10)$, and 5 three – dimensional faces $(f_3 = 5)$.

The number of elements of different dimensions in this boundary complex coincide with the values of these quantities in the polytope 4 – simplex. Therefore, can to assume that the dimension of a tetrahedron with center is 4.

The incidence of the elements of the boundary complex in this case coincide with the values of these quantities in the polytope 4 – simplex too

$$k_{a_j}(4) = 4, j = 0 \div 5; k_{e_j}(4) = 3, j = 1 \div 10; k_{s_j}(4) = 2, j = 1 \div 10; k_{v_j}(4) = 1, j = 1 \div 5.$$

Therefore, can to assert that a tetrahedron with a center is topologically equivalent to a polytope 4 - simplex. We have previously proved that the polytope 4 - simplex has interior points. Consequently, a tetrahedron with its center has interior points. The boundary complex in it does not occupy this figure entirely. An indicator of the existence of internal points in a polytope of higher dimension can be the homogeneity of elements of different dimensions in the boundary complex. In a cube with a center and an octahedron centered, this homogeneity was broken, vertices appeared with different incidence values. This indicated that it is not impossible to deform continuously the figure to obtain a uniform distribution of vertices with the same incidence values, in which there may be interior points.

CONCLUSION

It is proved that the coefficient incidence of an element of dimension i of the boundary complex of an n - cube and n - simplex is equal to the difference of the dimension of the cube or simplex n and the dimension of this element i. The construction of n – cross – polytopes of various dimensions, including large dimension, revealed that the incidence coefficient of elements of this polytope is substantially higher than the incidence coefficient of the elements of the n – cubes and n – simplexes for equal values of the dimension of polytopes. The incidence coefficients of all three types of polytopes increase linearly with increasing n, but the proportionality coefficient of this dependence in cross polytopes is twice as large as the proportionality coefficient of cubes and simplexes.

Significantly larger values of the incidence coefficients of the elements of the boundary complex of n – cross - polytopes as compared to the n - cubes and n - simplexes should be taken into account in the study of structures involving the n – cross - polytopes.

The mathematical representation of the polytope boundary complex of higher dimension corresponds to the physical representation of the surface (Adam, 1938) molecular formation (cluster). All atoms of molecular formation are in the boundary complex of polytope. Moreover, two types of polytope of higher dimension, corresponding to molecular formation, are possible. The first type contains within itself a region of internal points free of atoms. The second type does not have internal points; the whole body of the polytope is a boundary complex. The region of interior points can be used to place atoms in it from the external environment. In this case, the polytope changes the type from the first to the second. All chemical interactions can occur with atoms of the boundary complex, in principle, in any place of their location.

REFERENCES

Adam, N. K. (1938). *The Physics and Chemistry of surfaces.* Oxford: Clarendon Press.

Coxeter, H.S.M. (1963). *Regular Polytopes.* New York: John Wiley & Sons, Inc.

Grunbaum, B. (1967). *Convex Polytopes.* London: Springer.

Lord, E. A., Mackay, A. L., & Ranganathan, S. (2006). *New Geometry for New Materials.* Cambridge: Cambridge University Press.

Poincare, A. (1895). Analysis situs. *J. de e` Ecole Polytechnique, 1,* 1 – 121.

Riemann, G. F. B. (1854). *On the Hypotheses Which Lie at the Foundations of Geometry.* Gottingen: Gottingen Observatory.

Stringham, W. I. (1880). Regular figures in n – dimensional space. *American Journal of Mathematics*, *3*(1), 1–14. doi:10.2307/2369441

Vasiliev, F. P., & Ivanitsky, A. Yu. (2008). *Linear programming*. Moscow: Factor Press.

Zhizhin, G. V. (2013). Relations for the number of faces of different dimensions in the tower of n - dimensional convex polytopes. *Materials (Basel)*, 9.

Zhizhin, G. V. (2014a). *World – 4D*. St. – Petersburg: Polytechnic Service.

Zhizhin, G. V. (2014b). On the higher dimension in nature. *Biosphere*, *6*(4), 313–318.

Zhizhin, G. V. (2017). *Chemical Compound Structures and the Higher Dimension of Molecules: Emerging Research and Opportunities*. Hershey, PA: IGI Global.

Zhizhin, G. V., & Diudea, M. V. (2017). Space of Nanoworld. In M. V. Putz & C. M. Marius (Eds.), *Sustainable nanosystems, development, properties, and applications* (pp. 221–245). Hershey, PA: IGI Global. doi:10.4018/978-1-5225-0492-4.ch007

KEY TERMS AND DIFINITIONS

Cluster: A chemical compound that is intermediate between a molecule and a volumetric body.

The Boundary Complex of Polytope With Dimension *n*: The set of faces of a polytope with dimension from zero to n - 1.

The Coefficient Incidence of an Element of the Boundary Complex: The number of elements of the boundary complex to which the given element of the boundary complex belongs, having a dimension one greater than this element.

The Set of Interior Points of the Polytope: The set of points belonging to a polytope, but not belonging to its boundary complex.

Chapter 3
The Number of Symmetry Transformation of Convex Regular Polytopes in the n – Space

ABSTRACT

The number of symmetry transformations of regular polytopes of dimension n (n - cubes, n - simplexes, n - cross polytopes) are considered, using symmetry transformation of their facets. In this chapter, it is investigated how a certain symmetry transformation of the facet leads to the transformation of the symmetry of the polytope, under the condition of continuity of the polytope as a whole. It is established that the number of symmetry transformations of a regular n - polytope is equal to the product of the number of symmetry transformations of the facet of the corresponding polytope by the number of facets in this polytope.

INTRODUCTION

In the study of symmetry transformations of three - dimensional convex regular polyhedrons, their symmetries are identified with the motions of these bodies, which lead to the alignment of the body with itself (Fedorov, 1885; Schläfli, 1901). In this case, a one-to-one mapping of the symmetry transformations (having a geometric meaning) is established with permutations of the vertices

DOI: 10.4018/978-1-5225-6968-8.ch003

of the polyhedron (having algebraic meaning). However, moving into a space of higher dimension, the visibility of symmetry transformations as a result of the motion of a polytope (a polyhedron of higher dimension) disappears. The possibility of extending a purely algebraic representation of the symmetry of convex bodies to a space of higher dimension remains unproven. In this connection, in this chapter, one proposes a new method for investigating the symmetry transformations of convex regular polytopes of higher dimension on the basis of a sequential study of the symmetry transformations of their facets (with allowance for their motion), provided the whole figure is continuous.

THE NUMBER OF SYMMETRY TRANSFORMATION OF REGULAR N – SIMPLEX

A regular simplex consists of five regular tetrahedrons (Stringham, 1880), whose faces are regular triangles. Therefore, the symmetry transformations of a regular simplex are linked inextricably with the symmetry transformations of a regular tetrahedron and a regular triangle as the faces of a regular simplex. The regular triangle (Figure 1) has six symmetry operations that reunite the triangle with itself:

3 mirror reflection operations with respect to the bisectors m, n, k of angles at the vertices a, b, c; 2 turns of the triangle relative to the center of the triangle, for example, clockwise to the angles $\frac{2\pi}{3}$, and $\frac{4\pi}{3}$; the identity transformation that leaves the triangle in place.

The remaining movements of the triangle, which combine it with oneself (rotations to angles multiple $\frac{2\pi}{3}$ clockwise and counterclockwise, combinations of rotations with mirror reflections) reduce to the symmetry transformations already listed. To each of the six symmetry transformations there corresponds a certain permutation of three vertices a, b, c. Since the number of these permutations is equal to 3!, the total number of these permutations is 6. It is essential that under the indicated symmetry transformations all elements of dimension less than the dimension of the triangle appearing in the triangle go over into one another or into themselves. The vertices pass into each other or into themselves, but also sides of triangle A, B, C cross one another or into themselves. In this case, the triangle does not break into its constituent parts.

Figure 1. A regular triangle

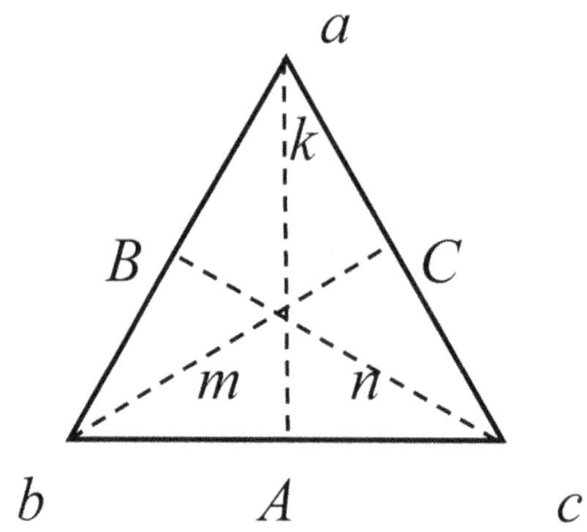

The number of permutation of the sides also equal to 3!. Moreover, to any permutation of sides there corresponds some permutation of vertices. For example, permutation of sides $\begin{pmatrix} ABC \\ BAC \end{pmatrix}$ corresponds to a permutation of vertices $\begin{pmatrix} abc \\ cba \end{pmatrix}$, and permutation of sides $\begin{pmatrix} ABC \\ BCA \end{pmatrix}$ corresponds to permutation of vertices $\begin{pmatrix} abc \\ bca \end{pmatrix}$.When the side is repainted by itself (the first case side *C*), it rotates 180 degrees.

Can to write that the total number of symmetry elements of the regular triangle is equal to 3! = 2!3, where 2! is the number of symmetry elements of the side of the triangle (reflection relative to the middle of the side, the identical transformation), and the second factor 3 is the number of elements of the triangle smaller by one unit of dimension in comparison with the triangle itself. This is reflected in the image of a triangle on a plane in the form of a polygon with three sides.

The tetrahedron has four symmetry axes l_1, l_2, l_3, l_4 of the third order, passing through the vertices 1, 2, 3, 4 and the centers of opposing faces (Figure 2).

Around each axis, except for the identical one, two more rotations are possible. They correspond such permutations

Figure 2. The regular tetrahedron

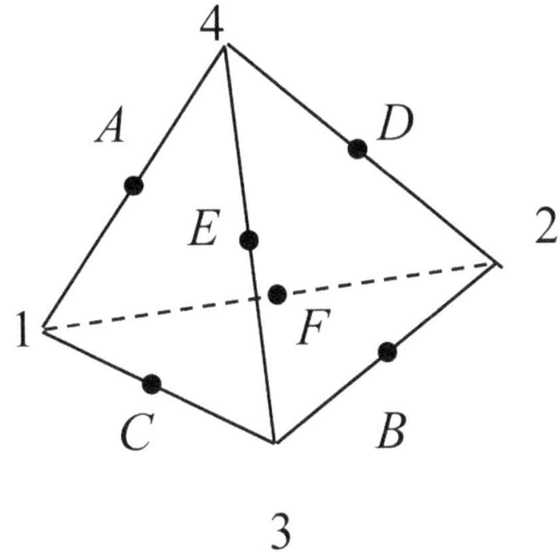

Around the axis l_1 : $\begin{pmatrix}1234\\1342\end{pmatrix}, \begin{pmatrix}1234\\1423\end{pmatrix}$,

Around the axis l_2 : $\begin{pmatrix}1234\\4213\end{pmatrix}, \begin{pmatrix}1234\\3241\end{pmatrix}$,

Around the axis l_3 : $\begin{pmatrix}1234\\2431\end{pmatrix}, \begin{pmatrix}1234\\4132\end{pmatrix}$,

Around the axis l_4 : $\begin{pmatrix}1234\\3124\end{pmatrix}, \begin{pmatrix}1234\\2314\end{pmatrix}$.

In addition, there are three axes second order passing through the middle of the crossing edges. Therefore, there are three more non - identical transformations

Around axis AB : $\begin{pmatrix}1234\\4321\end{pmatrix}$,

Around axis CD : $\begin{pmatrix}1234\\3412\end{pmatrix}$,

Around axis EF : $\begin{pmatrix}1234\\2143\end{pmatrix}$.

Thus, together with the identity transformation can to obtain 12 permutations. This is the group of tetrahedron rotations. The tetrahedron has 6 planes of symmetry, each of which passes through one of its edges and the middle of opposite edge. Can to choose one of the symmetry transformations and multiply it by 12 rotations group of the tetrahedron. Then can to get 12 more symmetry transformations of the tetrahedron. Together with the rotation group, these transformations constitute a group of symmetry transformations of the regular tetrahedron included of 24 transformations. Other symmetry transformations do not have a regular tetrahedron (Ponarin, 2015).

Each symmetry transformation of a tetrahedron reflected elements of dimension less than three: vertices to vertices, edges to edges, flat faces to flat faces. Each permutation of vertices corresponds to a completely definite permutation of its faces and a corresponding symmetry transformation of the tetrahedron. However, not every permutation of the edges of a tetrahedron corresponds to any symmetry transformation of the tetrahedron, if only because the number of edges of the tetrahedron (6) is greater than the number of vertices of the tetrahedron and the number of permutations of edges is much larger than the number of permutations of the vertices. There are permutations of edges that violate the continuity of the tetrahedron. Under symmetry transformations, the planar faces are mapped not only to each other, but also to themselves. For example, when reflect relative to the symmetry plane *AMD* (Figure 3), the planar faces $F_1 = ABD, F_2 = ACD$ are mapped to each other, and the flat faces $F_3 = ABC, F_4 = BCD$ are mapped on to themselves. When converting the symmetry of rotation to any third - order axis, the side faces that have a common vertex on this axis are mapped onto each other. Mirror reflections of the tetrahedron give rise to mirror reflections of its plane faces.

While the face perpendicular to the axis is mapped onto itself. Moreover, by virtue of the third order of the axis, the mapping onto itself is repeated twice and these mappings differ from the identity map.

The total number of symmetry elements of a regular tetrahedron can be expressed in the form $4! = 3!4$, where $3!$ is the number of symmetry elements of the plane side of the tetrahedron, and the number 4 is the number of sides of the tetrahedron, i.e. the number of elements of a tetrahedron with dimension one less than the dimension of a tetrahedron.

Now can to procced with describe the symmetry transformations of a regular 4 – simplex. Here, as in the description of the symmetry transformations of a triangle and a tetrahedron, there shall rely on the graphic representation

Figure 3. Plane of symmetry (MAD) in a regular tetrahedron

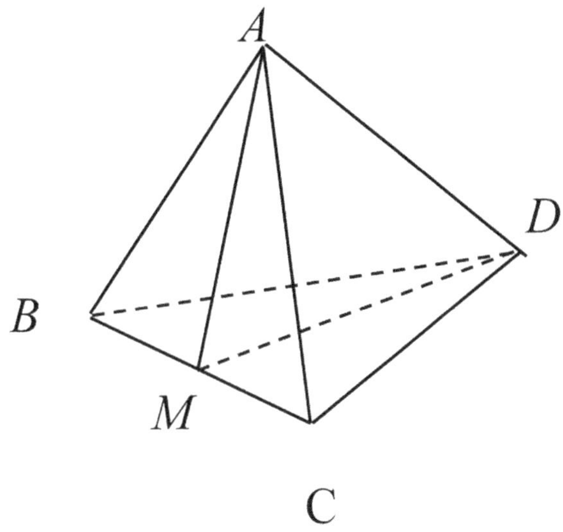

of a simplex, considering the symmetry transformations as geometric transformations that combine the image of the simplex with itself as result of the motion of the simplex as a whole. There will follow in it of the ideas of Fedorov, developed by him in the study of groups of symmetry transformations of geometric figures (Fedorov, 1885).

Let there be a regular simplex. There will consider the transformations of a simplex using symmetry transformations of tetrahedrons. Can to denote the vertices of the simplex in the initial state by the numbers 1, 2, ..., 5 (Figure 4). Can to denote this initial state t_0.

There is consideration of the transformation of a simplex due to the rotation of tetrahedrons around the axes passing through the vertices of tetrahedrons. For example, there choose vertex 1 and tetrahedron 1234. One rotate this tetrahedron, for example clockwise, around an axis passing through vertex 1 and the center of the opposing triangle 234 by an angle of $\frac{2\pi}{3}$. The vertices of triangle 234 with this rotation will change their numbers also clockwise 423. Can once again to turn the tetrahedron by an angle of $\frac{2\pi}{3}$ around this axis (in total there will be a turn by an angle of $\frac{4\pi}{3}$). Then the numbering of the vertices will change again 342. The next turn will return the triangle

Figure 4. The polytope 4 - simplex

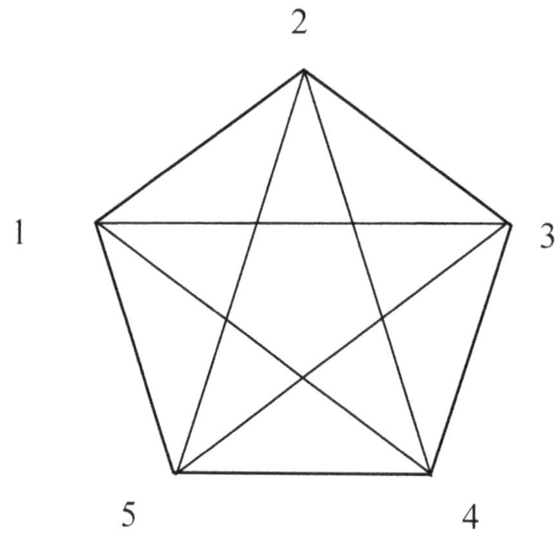

to its original state. The vertices 1 and 5 of the tetrahedron remain in place at these turns.

It is obvious that when the tetrahedron 2345 rotates about the axis passing through the vertex 5 and the vertex 5 and the vertex 1 remain in place. Moreover, the triangle 234 changes the vertex numbers as well as when the tetrahedron 1234 rotates. When a tetrahedron in a 4 - simplex rotates around the axis passing through the vertex of this tetrahedron, the number of vertices and other tetrahedrons that go into the 4 - simplex changes to preserve the continuity of the 4 -simplex as a whole. In this case, the common faces of neighboring tetrahedrons are conserved. This leads to a change in the order of the vertex numbers of the 4-simplex in comparison with the state of t_0.

In order to isolate the states of a 4 - simplex that are not reducible to each other by a cyclic permutation of vertices, we take as the first step the states that differ from one another by cyclic permutation of vertices, be considered identic.

Can to denote δ_{ij} the operation of rotation of tetrahedrons by an angle $\frac{2\pi}{3}$ around an axis passing through the vertices *i* or *j*. The vertices *i*, *j* remain in place with these rotations, the remaining vertices rotate clockwise. There shall denote δ_{ijf} such operations by an angle $\frac{4\pi}{3}$, in parentheses, indicate state of the 4 - simplex to which this operation is applied, for example

$\delta_{ij}(t_0)$ or $\delta_{ijf}(t_0)$ denote operations δ_{ij} and δ_{ijf} to the initial state. The remaining states will be denoted by the digits 1, 2, ... of the symbol *t*. The question arises how many such different states t_k can exist. Their number is equal to the number of considered symmetry transformations of the 4 - simplex. One also introduce the operation of the product of rotations of tetrahedrons, denoting it $\delta_{ij}\delta_{lm}$. Let's agree that the first turn is in this record leftmost. The next turn is after it on the right. Let's consider operation δ_{12f}. It is easy to see that the new state of the 4 - simplex after the application of this operation is $12453 = t_1$. Then, $\delta_{12f}(t_0) = t_1$. One can find that another operation leads to this state $\delta_{45}(t_0) = t_1$. Thus, there are equalities $\delta_{45}(t_0) = \delta_{12f}(t_0) = t_1 = 12453$.

The following equalities are established in a similar way:

$$\delta_{24}(t_0) = \delta_{13f}(t_0) = t_2 = 14352,$$

$$\delta_{12}(t_0) = \delta_{34f}(t_0) = t_3 = 12534,$$

$$\delta_{23}(t_0) = \delta_{45f}(t_0) = t_4 = 52314,$$

$$\delta_{34}(t_0) = \delta_{15f}(t_0) = t_5 = 13425,$$

$$\delta_{35}(t_0) = \delta_{24f}(t_0) = t_6 = 41325,$$

$$\delta_{15}(t_0) = \delta_{23f}(t_0) = t_7 = 42351,$$

$$\delta_{13}(t_0) = \delta_{25f}(t_0) = t_8 = 32415,$$

$$\delta_{14}(t_0) = \delta_{35f}(t_0) = t_9 = 24315,$$

$$\delta_{25}(t_0) = \delta_{14f}(t_0) = t_{10} = 13542.$$

Thus, there are only 10 different states of the 4 - simplex after applying a one - time rotation operation of tetrahedrons from the initial state $t_0 = 12345, k = 10$. To each of the new states we will use operations δ_{ij} and δ_{ijf} to find other possible states of the 4 - simplex obtained by rotation of tetrahedrons entering into it. These states will correspond to the result of applying the product of rotation operations of tetrahedrons from the initial state t_0.

Let`s apply operations δ_{ij} and δ_{ijf} to the state $t_1 = 12453$

$$\delta_{12}(t_1) = 12345 = \delta_{45f}(t_1) = t_0,$$

$$\delta_{12f}(t_1) = 12534 = \delta_{35}(t_1) = t_3,$$

$$\delta_{24}(t_1) = 32415 = \delta_{35f}(t_1) = t_8,$$

$$\delta_{24f}(t_1) = 52431 = \delta_{13}(t_1) = t_9,$$

$$\delta_{15}(t_1) = 25413 = \delta_{34f}(t_1) = t_6,$$

$$\delta_{45}(t_1) = 31452 = \delta_{13f}(t_1) = t_4,$$

$$\delta_{23}(t_1) = 52143 = \delta_{15f}(t_1) = t_2,$$

$$\delta_{14}(t_1) = 42513 = \delta_{23f}(t_1) = t_5,$$

$$\delta_{34}(t_1) = 51423 = \delta_{25f}(t_1) = t_7,$$

$$\delta_{25}(t_1) = 15432 = \delta_{14f}(t_1) = t_{11}.$$

Thus, one (t_{11}) more state of the 4 - simplex was discovered in addition to the already defined 10 states. Most of the operations on the state t_1 are reduced to the previously considered operations on the state t_0.

Let's apply operations δ_{ij} and δ_{ijf} to the state $t_2 = 14352$

$$\delta_{12}\left(t_2\right) = \delta_{34f}\left(t_2\right) = t_{11},$$

$$\delta_{34}\left(t_2\right) = \delta_{25f}\left(t_2\right) = t_9,$$

$$\delta_{15}\left(t_2\right) = \delta_{23f}\left(t_2\right) = t_1,$$

$$\delta_{14}\left(t_2\right) = \delta_{15f}\left(t_2\right) = t_7,$$

$$\delta_{25}\left(t_2\right) = \delta_{14f}\left(t_2\right) = t_4,$$

$$\delta_{23}\left(t_2\right) = \delta_{45f}\left(t_2\right) = t_5,$$

$$\delta_{35}\left(t_2\right) = \delta_{12f}\left(t_2\right) = t_{10},$$

$$\delta_{13}\left(t_2\right) = \delta_{24f}\left(t_2\right) = t_0,$$

$$\delta_{24}\left(t_2\right) = \delta_{15f}\left(t_2\right) = t_6,$$

$$\delta_{45}\left(t_2\right) = \delta_{13f}\left(t_2\right) = t_8.$$

Can to see that all states of the 4 - simple formed from the application of rotation operations of tetrahedrons to state t_2 are already known 12 states, including the initial state t_0. It can be shown that the application of rotation operations of tetrahedrons to all 12 states of 4 – simplex leads to the same 12 states. In particular, the application of rotation operations of tetrahedrons to the state t_{11} of the 4 - simplex leads to transitions of states

$$\delta_{12}\left(t_{11}\right) = \delta_{45f}\left(t_{11}\right) = t_{10},$$

$$\delta_{34}\left(t_{11}\right) = \delta_{12f}\left(t_{11}\right) = t_2,$$

$$\delta_{15}\left(t_{11}\right) = \delta_{34f}\left(t_{11}\right) = t_9,$$

$$\delta_{14}\left(t_{11}\right) = \delta_{25f}\left(t_{11}\right) = t_1,$$

$$\delta_{25}\left(t_{11}\right) = \delta_{13f}\left(t_{11}\right) = t_7,$$

$$\delta_{23}\left(t_{11}\right) = \delta_{15f}\left(t_{11}\right) = t_8,$$

$$\delta_{35}\left(t_{11}\right) = \delta_{14f}\left(t_{11}\right) = t_5,$$

$$\delta_{13}\left(t_{11}\right) = \delta_{24f}\left(t_{11}\right) = t_3,$$

$$\delta_{24}\left(t_{11}\right) = \delta_{35f}\left(t_{11}\right) = t_4,$$

$$\delta_{45}\left(t_{11}\right) = \delta_{23f}\left(t_{11}\right) = t_6.$$

Consider transformations of a 4 - simplex as a result of the rotation of tetrahedrons around a line joining the middle of the opposite edges. In the initial state $t_0 = 12345$ there are five tetrahedrons

$T_1 = 1234, T_2 = 2345, T_3 = 1345, T_4 = 1245, T_5 = 1235$. In the tetrahedron T_1 there are pairs of the opposite edges 23 – 14, 12 – 34, 24 – 13. When rotating of T_1 around the lines connecting the middle edges of these pairs, the vertices in the edges change places. Therefore, the rotation of the tetrahedron T_1 around these lines leads to changes in the initial state t_0 of the 4 - simplex

$$\delta_{23-14}(t_0) = 43215 = t_{11}, \delta_{12-34}(t_0) = 21435 = t_2, \delta_{24-13}(t_0) = 34125 = t_3.$$

In the tetrahedron T_2 there are pairs of the opposite edges 25 – 34, 23 – 54, 24 – 35. When rotating of T_2 around the lines connecting the middle edges of these pairs, the vertices in the edges change places. Therefore, the rotation of the tetrahedron T_2 around these lines leads to changes in the initial state t_0 of the 4 - simplex

$$\delta_{25-34}(t_0) = 15432 = t_{11}, \delta_{23-54}(t_0) = 13254 = t_6, \delta_{24-35}(t_0) = 14523 = t_4.$$

In the tetrahedron T_3 there are pairs of the opposite edges 13 – 54, 15 – 34, 14 – 35. When rotating of T_3 around the lines connecting the middle edges of these pairs, the vertices in the edges change places. Therefore, the rotation of the tetrahedron T_3 around these lines leads to changes in the initial state t_0 of the 4 - simplex

$$\delta_{13-54}(t_0) = 32154 = t_{11}, \delta_{15-34}(t_0) = 52431 = t_9, \delta_{14-35}(t_0) = 42513 = t_5.$$

In the tetrahedron T_4 there are pairs of the opposite edges 12 – 45, 15 – 24, 14 – 25. When rotating of T_4 around the lines connecting the middle edges of these pairs, the vertices in the edges change places. Therefore, the rotation of the tetrahedron T_4 around these lines leads to changes in the initial state t_0 of the 4 - simplex

$$\delta_{12-45}(t_0) = 21354 = t_{10}, \delta_{15-24}(t_0) = 54321 = t_{11}, \delta_{14-25}(t_0) = 45312 = t_1.$$

In the tetrahedron T_5 there are pairs of the opposite edges 15 – 23, 12 – 34, 13 – 25. When rotating of T_5 around the lines connecting the middle edges of these pairs, the vertices in the edges change places. Therefore, the rotation of the tetrahedron T_5 around these lines leads to changes in the initial state t_0 of the 4 - simplex

$$\delta_{15-23}(t_0) = 53241 = t_8, \delta_{12-35}(t_0) = 21543 = t_{11}, \delta_{13-25}(t_0) = 35142 = t_7.$$

Thus, can to see that the rotations of tetrahedrons around the lines connecting the middle of the opposite edges from the initial state t_0 lead to the already known states $t_1 \div t_{11}$ of the 4 -simplex. It is necessary to investigate the changes of states under such rotations and from other states of the 4 - simplex.

One choose the permissive state of the 4 - simplex, for example, t_9, in this state there are five tetrahedrons $T_6 = 1524, T_7 = 5243, T_8 = 1543, T_9 = 1243, T_{10} = 1523$.

In the tetrahedron T_6 there are pairs of the opposite edges 14 – 25, 15 – 24, 12 – 45. When rotating of T_6 around the lines connecting the middle edges of these pairs, the vertices in the edges change places. Therefore, the rotation of the tetrahedron T_6 around these lines leads to changes in the initial state t_9 of the 4 - simplex

$$\delta_{14-25}(t_9) = 51342 = t_5, \delta_{15-24}(t_9) = 42351 = t_7, \delta_{12-45}(t_9) = 15324 = t_8.$$

In the tetrahedron T_7 there are pairs of the opposite edges 25 – 43, 53 – 24, 23 – 45. When rotating of T_7 around the lines connecting the middle edges of these pairs, the vertices in the edges change places. Therefore, the rotation of the tetrahedron T_7 around these lines leads to changes in the initial state t_9 of the 4 - simplex

$$\delta_{25-43}(t_9) = 53412 = t_3, \delta_{53-24}(t_9) = 42513 = t_5, \delta_{23-45}(t_9) = 35214 = t_2.$$

In the tetrahedron T_8 there are pairs of the opposite edges 15 – 43, 13 – 45, 14 – 35. When rotating of T_8 around the lines connecting the middle edges of these pairs, the vertices in the edges change places. Therefore, the rotation of the tetrahedron T_8 around these lines leads to changes in the initial state t_9 of the 4 - simplex

$$\delta_{15-43}(t_9) = 23451 = t_0, \delta_{13-45}(t_9) = 25134 = t_5, \delta_{14-35}(t_9) = 21543 = t_{11}.$$

In the tetrahedron T_9 there are pairs of the opposite edges 12 – 43, 24 – 31, 14 – 23. When rotating of T_9 around the lines connecting the middle edges of these pairs, the vertices in the edges change places. Therefore, the rotation of the tetrahedron T_9 around these lines leads to changes in the initial state t_9 of the 4 - simplex

$$\delta_{12-43}(t_9) = 13425 = t_5, \delta_{24-31}(t_9) = 42135 = t_{10}, \delta_{14-23}(t_9) = 31245 = t_1.$$

In the tetrahedron T_{10} there are pairs of the opposite edges 15 – 23, 12 – 34, 13 – 25. When rotating of T_{10} around the lines connecting the middle edges of these pairs, the vertices in the edges change places. Therefore, the rotation of the tetrahedron T_{10} around these lines leads to changes in the initial state t_9 of the 4 - simplex

$$\delta_{15-23}(t_9) = 34251 = t_5, \delta_{13-25}(t_9) = 54132 = t_6, \delta_{12-35}(t_9) = 14523 = t_4.$$

Can to see that all rotations of the tetrahedrons from the state t_9 of the 4 - simplex lead to the already known states of this polytope. It can be verified that the indicated rotations of tetrahedrons in other states of the 4 - simplex lead all to the same already known states of this polytope.

It is interesting that the tetrahedron rotation group containing 12 symmetry transformations induced the symmetry transformation group of the boundary complex of the 4 - simplex, which also contains 12 symmetry transformations. In this case, the rotations of tetrahedrons around the axes connecting the vertices of tetrahedrons and the axes connecting the midpoints of the adjacent edges lead to the same elements of the symmetry transformation group of the 4 - simplex boundary complex.

Consider transformations of a 4 - simplex under mirror reflections of tetrahedrons. Let there be a 4 - simplex in the initial state t_0. Let us consider a tetrahedron $T_1 = 1234$. One reflect it with respect to the plane passing through the edge 23 perpendicular to the plane of the triangle 134. Then the vertices 1 and 4 change places $s_{23}\left(t_0\right) = 42315 = t_{0,1}$. In this case, other tetrahedrons in the 4 - simplex from the continuity condition of the 4 - simplex also undergo a mirror transformation. Indeed, for example, prior to the transformation of the tetrahedron 1234, the tetrahedron 1234 has been associated with the tetrahedron 1245 along the triangle 124. After the transformation, the triangle 241 is located at the tetrahedron 1245. This triangle is followed by the tetrahedron 4215, which is connected with the tetrahedron 1245 by the mirror reflection along the edge 25. Analogous transformations also occur with other tetrahedrons that come to contact with tetrahedron 1234 along other triangles. For example, the tetrahedron 1345 was in contact with the tetrahedron 1234 along the triangle 134. After a mirror transformation of the symmetry of the tetrahedron 1234, a triangle 431 becomes common between the two tetrahedrons. At that, the tetrahedron

1345 is transformed into a tetrahedron 4315, which is the result of the mirror transformation of the tetrahedron 1345 along the edge 35 (and so further).

Thus, the tetrahedron $T_1 = 1234$ has 6 mirror transformations on 6 edges

$$s_{34}\left(T_1\right) = 2134, s_{24}\left(T_1\right) = 3214, s_{23}\left(T_1\right) = 4231, s_{14}\left(T_1\right) = 1324, s_{12}\left(T_1\right) = 1243, s_{13}\left(T_1\right) = 1432.$$

This leads to six changes in the state t_0 of the 4 - simplex

$$s_{34T_1}\left(t_0\right) = 21345 = t_{0,1}, s_{24T_1}\left(t_0\right) = 32145 = t_{0,2}, s_{23T_1}\left(t_0\right) = 42315 = t_{0,3},$$

$$s_{14T_1}\left(t_0\right) = 13245 = t_{0,4}, s_{12T_1}\left(t_0\right) = 12435 = t_{0,5}, s_{13T_1}\left(t_0\right) = 14325 = t_{0,6}.$$

The tetrahedron $T_5 = 1235$ has 6 mirror transformations on 6 edges

$$s_{35}\left(T_5\right) = 2135, s_{13}\left(T_5\right) = 1532, s_{15}\left(T_5\right) = 1325, s_{52}\left(T_5\right) = 3215, s_{32}\left(T_5\right) = 5231, s_{12}\left(T_5\right) = 1253.$$

This leads to six changes in the state t_0 of the 4 - simplex

$$s_{35T_5}\left(t_0\right) = 21345 = t_{0,1}, s_{13T_5}\left(t_0\right) = 15342 = t_{0,2}, s_{15T_5}\left(t_0\right) = 13245 = t_{0,4},$$

$$s_{14T_1}\left(t_0\right) = 13245 = t_{0,4}, s_{12T_1}\left(t_0\right) = 12435 = t_{0,5}, s_{13T_1}\left(t_0\right) = 14325 = t_{0,6}.$$

The tetrahedron $T_4 = 1245$ has 6 mirror transformations on 6 edges

$$s_{12}\left(T_4\right) = 1254, s_{14}\left(T_4\right) = 1542, s_{15}\left(T_4\right) = 1425, s_{24}\left(T_4\right) = 5241, s_{25}\left(T_4\right) = 4215, s_{45}\left(T_4\right) = 2145.$$

This leads to six changes in the state t_0 of the 4 - simplex

$$s_{12T_4}\left(t_0\right) = 12354 = t_{0,10}, s_{14T_4}\left(t_0\right) = 15342 = t_{0,7}, s_{15T_4}\left(t_0\right) = 14325 = t_{0,6},$$

$$s_{24T_4}\left(t_0\right) = 52341 = t_{0,8}, s_{25T_4}\left(t_0\right) = 42315 = t_{0,3}, s_{45T_4}\left(t_0\right) = 21345 = t_{0,1}.$$

It can be verified that mirror transformations of tetrahedrons $T_2 = 2345, T_3 = 1345$ lead to the already known mirror transformations of the 4 - simplex. Thus, the mirror transformations of the tetrahedrons entering the 4 - simplex state are brought to the new 10 states of the 4 - simplex $t_{0,1} \div t_{0,10}$.

Let there be a 4 - simplex in the state $t_1 =$ 12453. In it there are five tetrahedrons

$$T_{11} = 1253, T_{12} = 1245, T_{13} = 2453, T_{14} = 1453, T_{15} = 1243.$$

It should be noted that the change in the number in the designation of a tetrahedron is associated with a change in the numbering of its vertices under symmetry transformations. The number of tetrahedrons in a 4 - simplex always naturally remains equal to five.

The tetrahedron $T_{11} = 1253$ has 6 mirror transformations on 6 edges

$$s_{12}\left(T_{11}\right) = 1235, s_{13}\left(T_{11}\right) = 1523, s_{15}\left(T_{11}\right) = 1352, s_{25}\left(T_{11}\right) = 3251, s_{53}\left(T_{11}\right) = 2153, s_{23}\left(T_{11}\right) = 5213.$$

This leads to six changes in the state t_1 of the 4 - simplex

$$s_{12T_{11}}\left(t_1\right) = 12435 = t_{0,5}, s_{13T_{11}}\left(t_1\right) = 15423 = t_{0,3}, s_{15T_{11}}\left(t_1\right) = 13452 = t_{0,1},$$

$$s_{25T_{11}}\left(t_1\right) = 32451 = t_{0,4}, s_{53T_{11}}\left(t_1\right) = 21453 = t_{0,2}, s_{23T_{11}}\left(t_1\right) = 52413 = t_{1,1}.$$

The tetrahedron $T_{12} = 1245$ has 6 mirror transformations on 6 edges

$$s_{12}\left(T_{12}\right) = 1254, s_{14}\left(T_{12}\right) = 1542, s_{25}\left(T_{12}\right) = 4215, s_{24}\left(T_{12}\right) = 5241, s_{45}\left(T_{12}\right) = 2145, s_{15}\left(T_{12}\right) = 1425.$$

This leads to six changes in the state t_1 of the 4 - simplex

$$s_{12T_{12}}\left(t_1\right) = 12543 = t_{0,9}, s_{14T_{12}}\left(t_1\right) = 15423 = t_{0,3}, s_{25T_{12}}\left(t_1\right) = 42153 = t_{0,7},$$

$$s_{24T_{12}}\left(t_1\right) = 52413 = t_{1,1}, s_{45T_{12}}\left(t_1\right) = 21453 = t_{0,2}, s_{15T_{12}}\left(t_1\right) = 14253 = t_{1,2}.$$

The tetrahedron $T_{13} = 2453$ has 6 mirror transformations on 6 edges

$$s_{24}\left(T_{13}\right) = 2435, s_{23}\left(T_{13}\right) = 2543, s_{25}\left(T_{13}\right) = 2354, s_{45}\left(T_{13}\right) = 3452, s_{43}\left(T_{13}\right) = 5423, s_{53}\left(T_{13}\right) = 4253.$$

This leads to six changes in the state t_1 of the 4 - simplex

$$s_{24T_{13}}\left(t_1\right) = 12435 = t_{0,5}, s_{23T_{13}}\left(t_1\right) = 12543 = t_{0,9}, s_{25T_{13}}\left(t_1\right) = 12354 = t_{0,10},$$

$$s_{45T_{13}}\left(t_1\right) = 13452 = t_{0,1}, s_{43T_{13}}\left(t_1\right) = 15423 = t_{0,3}, s_{53T_{13}}\left(t_1\right) = 14253 = t_{1,2}.$$

The tetrahedron $T_{14} = 1453$ has 6 mirror transformations on 6 edges

$$s_{14}\left(T_{14}\right) = 1435, s_{13}\left(T_{14}\right) = 1543, s_{45}\left(T_{14}\right) = 3451, s_{15}\left(T_{14}\right) = 1354, s_{43}\left(T_{14}\right) = 5413, s_{53}\left(T_{14}\right) = 4153.$$

This leads to six changes in the state t_1 of the 4 - simplex

$$s_{14T_{14}}\left(t_1\right) = 12435 = t_{0,5}, s_{13T_{14}}\left(t_1\right) = 12543 = t_{0,9}, s_{45T_{14}}\left(t_1\right) = 32451 = t_{0,4},$$

$$s_{15T_{14}}\left(t_1\right) = 12354 = t_{0,10}, s_{43T_{14}}\left(t_1\right) = 52413 = t_{1,1}, s_{53T_{14}}\left(t_1\right) = 42153 = t_{0,7}.$$

The tetrahedron $T_{15} = 1243$ has 6 mirror transformations on 6 edges

$$s_{12}\left(T_{15}\right) = 1234, s_{14}\left(T_{15}\right) = 1342, s_{13}\left(T_{15}\right) = 1423, s_{24}\left(T_{15}\right) = 3241, s_{23}\left(T_{15}\right) = 4213, s_{43}\left(T_{15}\right) = 2143.$$

This leads to six changes in the state t_1 of the 4 - simplex

$$s_{12T_{15}}\left(t_1\right) = 12354 = t_{0,10}, s_{14T_{15}}\left(t_1\right) = 13452 = t_{0,1}, s_{13T_{15}}\left(t_1\right) = 14253 = t_{1,2},$$

$$s_{24T_{15}}\left(t_1\right) = 32451 = t_{0,4}, s_{23T_{15}}\left(t_1\right) = 42153 = t_{0,7}, s_{43T_{15}}\left(t_1\right) = 21453 = t_{0,7}.$$

Thus, mirror reflections of tetrahedrons lead to 12 states of the boundary complex of the 4 -simplex from states t_0, and t_1. These are the states $t_{0,1} \div t_{0,10}, t_{1,1}, t_{1,2}$. It can be verified that the mirror reflections of tetrahedrons from any states of the 4-simplex lead again to one of the states $t_{0,1} \div t_{0,10}, t_{1,1}, t_{1,2}$ of the 4 - simplex. Any non - trivial combinations of the turns of tetrahedrons and their mirror reflections also lead to one of the states $t_{0,1} \div t_{0,10}, t_{1,1}, t_{1,2}$ of the 4 - simplex.

Consequently, the boundary complex of the 4 - simplex has, without taking into account the possible cyclic permutation of the vertices, of only 24 states. 12 of them arise from the rotation of tetrahedrons and 12 more are formed as a result of mirror reflections of tetrahedrons and combinations of mirror reflections and rotations. If we now consider the possibility of cyclic permutation of vertices 4 - simplex, each of the 24 states gives rise to 5 different states of the boundary complex of the 4 - simplex. Then the total number of elements of the symmetry transformations of a 4 - complex is $4!\cdot 5 = 120$. Here the first factor 4! = 24 is the number of states of the 4 - simplex due to rotations and mirror reflections of tetrahedrons. One can assume that these states it are induced by the group of symmetry transformations of the tetrahedron, and the number of these states is equal to the number of symmetry transformations of the tetrahedron.

Thus, for all simplexes of dimension n considered for $n = 2, 3, 4$ there is one law: the number of symmetry elements of a simplex of dimension n is equal to the product of the number of symmetry elements of a simplex of dimension n - 1 by the number of vertices of a simplex of dimension n. It can be formulated as follows: the number of symmetry elements of a regular n -simplex is equal to the product of the number of symmetry elements of its facets by the number of facets. The equality of the number of vertices and the number of facets in a regular n - simplex follows from the formula for the number of faces of dimension j in the n - simplex (Zhizhin, 2014)

$$f_j(n) = C_{n+1}^{j+1}. \tag{1}$$

Substituting into the formula (1) equality $j = n - 1$, can to obtain

$$f_{n-1}(n) = C_{n+1}^{n} = n + 1 = C_{n+1}^{1} = f_0(\mathrm{n}) \tag{2}$$

If one denote the number of symmetry elements of an n - simplex $N_{s,n}$, then this assertion can be written in the form of formula

$$N_{s,n} = N_{s,n-1} f_{n-1}. \tag{3}$$

Substituting the value f_{n-1} from 2 into 3, can to obtain

$$N_{s,n} = N_{s,n-1} f_{n-1} = N_{s,n-1}(n + 1). \tag{4}$$

Given the recurrence of the formula (4), it follows from this that

$$N_{s,n} = (\mathrm{n}+1)!. \tag{5}$$

The last expression coincides with the formula obtained by Rosenfeld (Rosenfeld, 1966) by the algebraic path. However, in this case, the geometric singularities of the boundary complexes of simplexes of higher dimension remained unexplored.

In projection on a two - dimensional plane the regular n - simplex has the form of a regular (n + 1) - angled, each vertex of which is connected by an edge with all the other vertices. The number of vertices in this polygon is equal to the number of facets of the corresponding n - simplex. For example, for a regular 5 - simplex with respect to the formula (2), there are

$$f_4(5) = C_{5+1}^{5} = 6 = C_6^1 = f_0(5).$$

The projection of 5 – simplex on a two - dimensional plane has form of a regular 6 – angled, each vertex of which is connected by an edge with all the other vertices (Figure 5).

In accordance with formula (3), the number of symmetry transformations of the regular 5 -simplex equal product of the number of transformations of the symmetry of the 4 - simplex by the number of facets of the 5 - simplex

$$N_{s,5} = N_{s,4} 6 = 120 \cdot 6 = 5! \cdot 6 = 6! = 720.$$

Figure 5. The polytope 5 - simplex

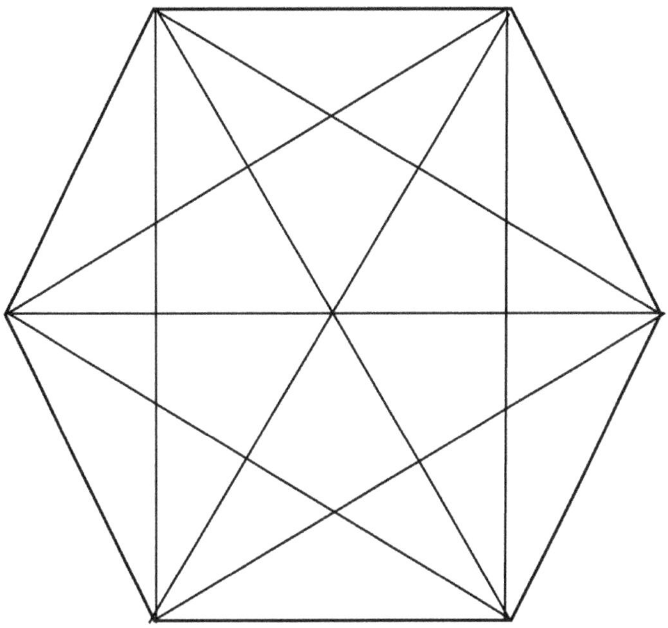

THE NUMBER OF SYMMETRY TRANSFORMATION OF N – CUBE

The symmetry transformations of a 4 – cube is linked inextricably with the symmetry transformations of a 3 – cube and a square as the faces of a 4 – cube. The square has two planes of symmetry, passing along the diagonals of the square and two planes of symmetry, passing through the centers of parallel sides of the square. In addition, the square has a symmetry axis of the fourth order, passing through the center of the square. These symmetry elements give rise to 8 independent elements of the symmetry transformations (Figure 6).

If one denote the vertices of the square *a, b, c, d,* then from 24 permutations of the vertices only 8 permutations correspond to independent symmetry transformations *abcd, badc, dcba, cbad, adcb, bcda, cdab, dabc.* For a square, can to write the number of symmetry elements as a product $2 \cdot 4$, where 2 is the number of symmetry elements of the side of the square, and 4 is the number of sides of the square.

The polytope 3 – cube has axis of different order.

Figure 6. The elements of the symmetry transformations of the square

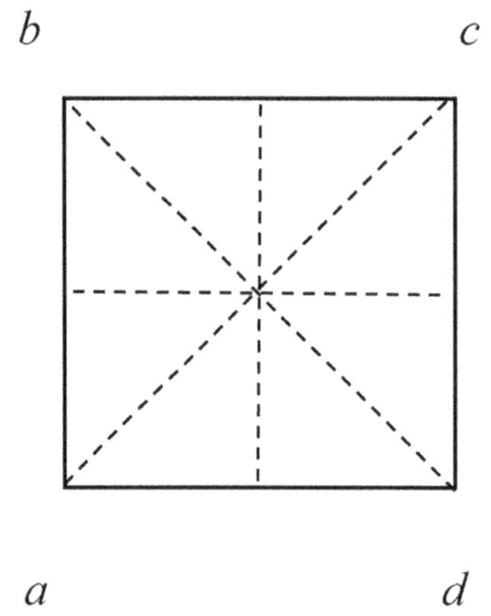

1. The tree axis passing through centers opposite faces have four order: $a_1 = o_1o_2, a_2 = o_3o_4, a_3 = o_5o_6$. Around each of these axes, non-identical rotations (clockwise) to corners $\alpha_1 = \frac{\pi}{2}, \alpha_2 = \pi, \alpha_3 = \frac{3\pi}{2}$ are possible. If the initial state of the 3 - cube corresponds to the notation in Figure 7, these rotations are denoted as follows

$$a_1(\alpha_1) = \begin{pmatrix} 12345678 \\ 23416785 \end{pmatrix}, a_1(\alpha_2) = \begin{pmatrix} 12345678 \\ 34127856 \end{pmatrix}, a_1(\alpha_3) = \begin{pmatrix} 12345678 \\ 41238567 \end{pmatrix},$$

$$a_2(\alpha_1) = \begin{pmatrix} 12345678 \\ 51486237 \end{pmatrix}, a_2(\alpha_2) = \begin{pmatrix} 12345678 \\ 65872143 \end{pmatrix}, a_2(\alpha_3) = \begin{pmatrix} 12345678 \\ 26731584 \end{pmatrix},$$

$$a_3(\alpha_1) = \begin{pmatrix} 12345678 \\ 56218734 \end{pmatrix}, a_3(\alpha_2) = \begin{pmatrix} 12345678 \\ 87654321 \end{pmatrix}, a_3(\alpha_3) = \begin{pmatrix} 12345678 \\ 43781265 \end{pmatrix}.$$

2. The four axis are diagonals of the 3 - cube:

Figure 7. The polytope 3 – cube

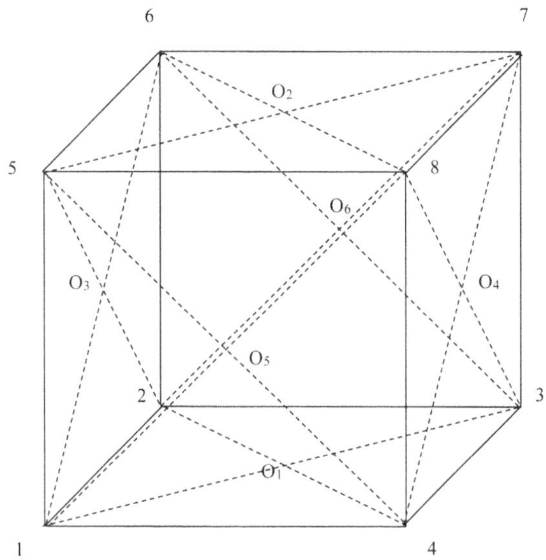

$b_1 = [1,7], b_2 = [2,8], b_3 = [3,5], b_4 = [4,6].$

This axis have third order. Around each of these axes, non - identical rotations to corners $\beta_1 = \frac{2\pi}{3}, \beta_2 = \frac{4\pi}{3}$ are possible

$$b_1(\beta_1) = \begin{pmatrix} 12345678 \\ 14852376 \end{pmatrix}, b_1(\beta_2) = \begin{pmatrix} 12345678 \\ 15624873 \end{pmatrix},$$

$$b_2(\beta_1) = \begin{pmatrix} 12345678 \\ 32674158 \end{pmatrix}, b_2(\beta_2) = \begin{pmatrix} 12345678 \\ 62157348 \end{pmatrix},$$

$$b_3(\beta_1) = \begin{pmatrix} 12345678 \\ 84375126 \end{pmatrix}, b_3(\beta_2) = \begin{pmatrix} 12345678 \\ 67325841 \end{pmatrix},$$

$$b_4(\beta_1) = \begin{pmatrix} 12345678 \\ 37842651 \end{pmatrix}, b_4(\beta_2) = \begin{pmatrix} 12345678 \\ 85147623 \end{pmatrix}.$$

3. The six axis of two order united the middles of the opposite edges:

$$c_1 = [1,2]-[7,8], c_2 = [5,6]-[4,3], c_3 = [6,7]-[1,4],$$
$$c_4 = [2,3]-[5,8], c_5 = [2,6]-[8,4], c_6 = [1,5]-[3,7].$$

Around each axis is one non-identical turn. This turns correspondent permutations

$$c_1 = \begin{pmatrix} 12345678 \\ 21563487 \end{pmatrix}, c_2 = \begin{pmatrix} 12345678 \\ 78436512 \end{pmatrix}, c_3 = \begin{pmatrix} 12345678 \\ 34127856 \end{pmatrix},$$
$$c_4 = \begin{pmatrix} 12345678 \\ 73268415 \end{pmatrix}, c_5 = \begin{pmatrix} 12345678 \\ 76583214 \end{pmatrix}, c_6 = \begin{pmatrix} 12345678 \\ 58761432 \end{pmatrix}.$$

The 3 - cube have five the planes of symmetry:

$$S_1(1573) = \begin{pmatrix} 12345678 \\ 14325876 \end{pmatrix}, S_2(2684) = \begin{pmatrix} 12345678 \\ 32147658 \end{pmatrix}, \mathrm{S}_3(o_1 o_2 o_5 o_6) = \begin{pmatrix} 12345678 \\ 43218765 \end{pmatrix},$$
$$S_4(o_1 o_2 o_3 o_4) = \begin{pmatrix} 12345678 \\ 21436587 \end{pmatrix}, S_5(o_3 o_4 o_5 o_6) = \begin{pmatrix} 12345678 \\ 56781234 \end{pmatrix}.$$

In addition, the 3 - cube has a center of symmetry that coincides with the center of the 3 - cube:

$$S = \begin{pmatrix} 12345678 \\ 78563412 \end{pmatrix}.$$

By composing different products of the listed symmetry transformations together with the trivial transformation *E*, one can find that apart from these 30 symmetry transformations, there are 18 more symmetry transformations. Together, 48 symmetry transformations of a 3 - cube form a group of symmetry transformations so that any product of these transformations leads to one of the symmetry transformations of this group. These 18 symmetry transformations have the form:

$$P_1 = \begin{pmatrix} 12345678 \\ 67852341 \end{pmatrix}, P_2 = \begin{pmatrix} 12345678 \\ 85674123 \end{pmatrix}, P_3 = \begin{pmatrix} 12345678 \\ 51268437 \end{pmatrix}, P_4 = \begin{pmatrix} 12345678 \\ 58416732 \end{pmatrix}, P_5 = \begin{pmatrix} 12345678 \\ 73486215 \end{pmatrix},$$
$$P_6 = \begin{pmatrix} 12345678 \\ 41583267 \end{pmatrix}, P_7 = \begin{pmatrix} 12345678 \\ 84157326 \end{pmatrix}, P_8 = \begin{pmatrix} 12345678 \\ 62375148 \end{pmatrix}, P_9 = \begin{pmatrix} 12345678 \\ 158442673 \end{pmatrix}, P_{10} = \begin{pmatrix} 12345678 \\ 37624851 \end{pmatrix},$$
$$P_{11} = \begin{pmatrix} 12345678 \\ 48731562 \end{pmatrix}, P_{12} = \begin{pmatrix} 12345678 \\ 23761485 \end{pmatrix}, P_{13} = \begin{pmatrix} 12345678 \\ 26513784 \end{pmatrix}, P_{14} = \begin{pmatrix} 12345678 \\ 76238514 \end{pmatrix}, P_{15} = \begin{pmatrix} 12345678 \\ 87345621 \end{pmatrix},$$
$$P_{16} = \begin{pmatrix} 12345678 \\ 12654378 \end{pmatrix}, P_{17} = \begin{pmatrix} 12345678 \\ 65127843 \end{pmatrix}, P_{18} = \begin{pmatrix} 12345678 \\ 34872156 \end{pmatrix}.$$

Thus, the number of the elements symmetry transformation of the 3 – cube also equal of the product the number elements symmetry transformation of the 2 – cube (square) by the number of the 2 – cube in the 3 – cube $8 \cdot 6 = 48$.

Now consider the 4 - cube. Let the vertices of a 4 - cube be designated as in Figure 8.

One of the eight 3 - cubes forming the 4 - cube boundary complex has a ground state 12345678. If you rotate this 3 - cube around the diagonal 1 - 7, then its vertex numbers will take values as in Figure 9.

In this case, the numbers of the remaining vertices of the 4 - cube, assuming 4 - cube is inseparable, will also change (see Figure 9), and, added to it unambiguously. Thus, a 4 - cube has as many states as there are 3 - cube states as result of its symmetry transformations, i.e. 48. If we now change the numbers of the vertices of the initial cube by a cyclic permutation, and such cyclic permutations 8, then each such changed state can be transformed by symmetry transformations 48 times. Thus, the total number of the symmetry transformations of a 4 - cube is 48 * 8 = 384. Thus, in the case of a 4 - cube, as before, the number of its symmetry transformations is equal to the product of the number of symmetry transformations facet by the facet number (equation (3)). However, the number of facets in the n - cube differs from the number of facets in the n - simplex. The number of faces of dimension i in the n - cube (Zhizhin, 2014) is

Figure 8. The 4 – cube in initial state

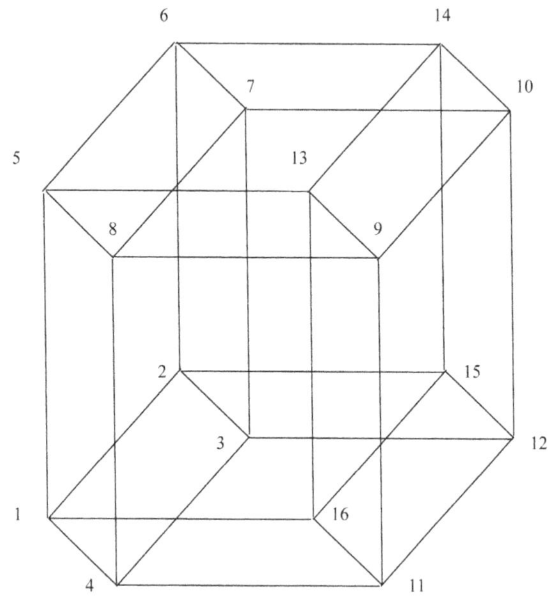

Figure 9. The 4 – cube after turn around diagonal 1-7

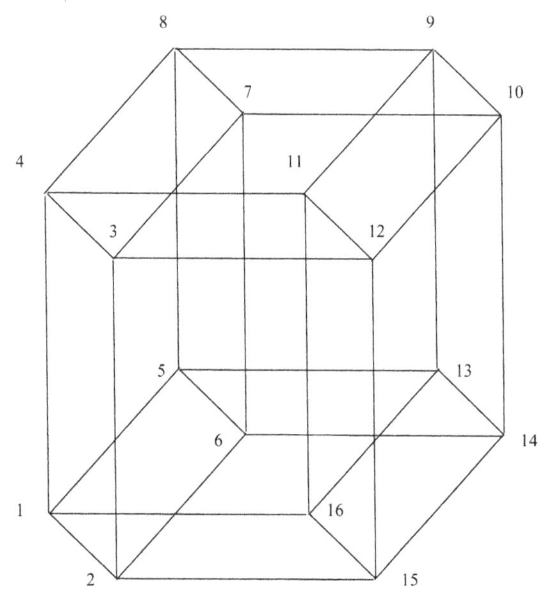

$$f_i = 2^{n-i} C_n^i. \tag{6}$$

Therefore, the number of facets in the n - cube ($i = n - 1$) is

$$f_{n-1} = 2 \cdot n. \tag{7}$$

Substituting the value f_{n-1} from (7) into (3), we obtain

$$N_{s,n} = N_{s,n-1} f_{n-1} = N_{s,n-1} 2n. \tag{8}$$

Given the recurrence of the formula (8), it follows from this that the number of the symmetry transformations of the n - cube

$$N_{s,n} = 2^n \text{ n!}. \tag{9}$$

Images of the 4 - cube in Figures 7 and 8 follow the work of Stringham (Stringham, 1880). One can suggest to represent the polytope 4 - cube in the projection on two-dimensional plasticity in the form of a polygon, similar to the image of a 4 - simplex. But in view of the large number of vertices, the polygon in this case contains vertices inside the outer contour (Figure 10). The change in the initial state of the 4 - cube in this case corresponds to the rotation of the projection of the 4 - cube in the plane around the center of the projection.

The number of the symmetry transformation of the 5 – cube concordantly of equations (7), (8), (9) equal $384 \cdot 10 = 2^5 \cdot 5! = 3840$.

One can also construct a symmetric image of the polytope 5 - cube on a two - dimensional plane (Figure 11).

It should be noted that as the dimension of the n - cube increases in its symmetrical projection, a superposition of a part of the edges and vertices occurs.

THE NUMBER OF SYMMETRY TRANSFORMATION OF N – CROSS - POLYTOPE

Symmetric images of the n – cross - polytopes with $n = 4, 5, 6$ were given in Chapter 2 (Figures 9, 10, 11). The distinctive feature of the n – cross –

Figure 10. The polytope 4 – cube in a symmetrical image

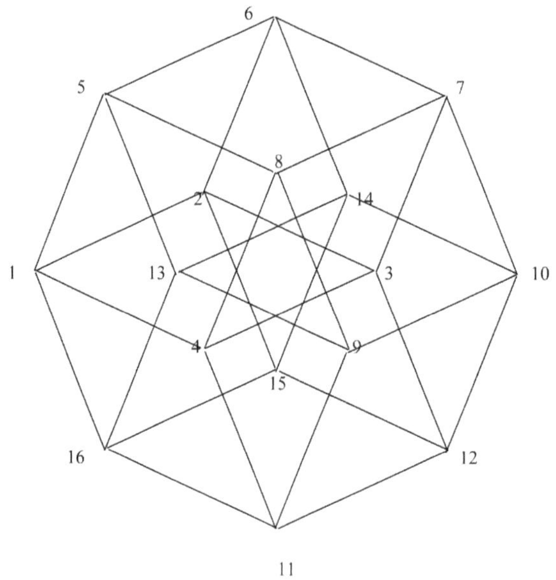

Figure 11. The polytope 5 – cube in a symmetrical image

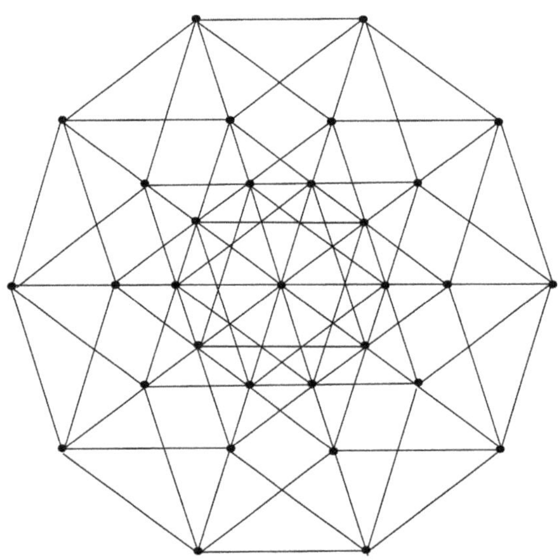

polytopes is that their opposite located vertices are not connected by an edge and their boundary complexes consist of tetrahedrons (Coxeter, 1962). For a n – cross - polytope, equation (3) is also valid, according to which the number of symmetry transformations is equal to the product of the number of facets by the number of their symmetry transformations. Since the n – cross - polytope is a simplicial polytope (Grunbaum, 1967; Ziegler, 1995), its facets are simplexes with dimension one less than the dimension of the cross - polytope. The number of facets for the n – cross - polytope is 2^n (Zhizhin, 2014). The number of symmetry transformations for a simplex of dimension n -1 is equal to $n!$, as follows from this chapter. Therefore, the number of symmetry transformations of the n – cross - polytope is $2^n \cdot n!$, as well as for the n - cube.

CONCLUSION

Thus, for regular convex polytopes of dimension n, a new law is established: the number of symmetry transformations of regular n - dimensional polytopes is equal to the product of the number of symmetry transformations of their facets by the number of facets in a given polytope.

The considered method for determining the number of symmetry transformations of convex regular polytopes of higher dimension on the basis of analysis of the symmetry transformations of their facets can be used to study symmetry transformations of convex semi - regular and irregular polytopes of higher dimension.

REFERENCES

Coxeter, H.S.M. (1963). *Regular Polytopes.* New York: John Wiley & Sons, Inc.

Fedorov, E. S. (1885). *The beginning of the doctrine of figures.* Saint - Petersburg: Academy of Sciences.

Grunbaum, B. (1967). *Convex Polytopes.* London: Springer.

Ponarin, Y. P. (2015). Elementary geometry (Vol. 2. *Stereometry, space transformation).* Moscow: MCNMO.

Rosenfeld, B. A. (1966). *Multidimensional spaces.* Moscow: Science.

Schläfli, L. (1901). *Theorie der vielfachen Kotinuität Zürcher und Furrer.*

Stringham, W. I. (1880). Regular figures in *n* - dimensional space. *American Journal of Mathematics*, *3*(1), 1–14. doi:10.2307/2369441

Zhizhin, G. V. (2014). *World – 4D*. St. – Petersburg: Polytechnic Service.

Ziegler, G. M. (1995). *Lectures on Polytopes*. Berlin: Springer – Verlag. doi:10.1007/978-1-4613-8431-1

KEY TERMS AND DIFINITIONS

C_n^m **:** Number of different combinations of *n* elements per *m*.

Facet: Face with dimension $n - 1$ in the polytope with dimension *n*.

N – Cross – Polytope: A convex polytope of dimension *n* in which every two vertices opposite to the center of the polytope have no connection by an edge, and the remaining vertices are joined by edges.

N – Cube: Convex polytope of dimension *n*, all vertices have connections with an edges to *n* other vertices of this polytope.

N – Simplex: Convex polytope of dimension *n*, all vertices of which have connections with an edge to all other vertices of the vertices of this polytope.

The Symmetry Transformation of the Figure: Figure movement, combining the figure with itself.

Chapter 4
Polytopic Prismahedrons

ABSTRACT

The structure of polytopes - polytopic prismahedrons, which are products of polytopes of lower dimensionality, is investigated. The products of polytopes do not belong to the well-studied class of simplicial polytopes, and therefore their investigations are of independent interest. Analytical dependencies characterizing the structure of the product of polytopes are obtained as a function of the structures of polytope factors. Images of a number of specific polytopic prismahedrons are obtained, tables of structures of polytopic prismahedrons are compiled, depending on the types of polytopes of the factors. The geometric properties of the boundary complexes of polytopic prismehedrons are investigated. Polytopic prismahedrons can be considered as a result of the chemical interaction of molecules, which, from among which there is a polytope of a certain dimension. The possibility of the multivaluedness of the incidence coefficients of geometric elements of polytopic prisms is revealed. It is shown that polytopic prismahedra, due to their nature, as products of polytopes, ensure the filling of n-dimensional spaces during their translation, creating a structure similar to the structure of quasicrystals.

THE DIFINITHION OF THE POLYTOPIC PRISMAHEDRONS

In the monograph G.V. Zhizhin (Zhizhin, 2017b) it was established that the fundamental area in the filling of the n - dimensional space is the polytopic prismahedron, which is a prism or a complex of prisms with bases in the form of polytopes of dimension n. It is polytopes provide the filling of the n -

DOI: 10.4018/978-1-5225-6968-8.ch004

dimensional space with a face into the face without gaps. Thus, the polytopic prismahedron is the sterehedron, the existence of which was suggested by B.N. Delone (Delone, 1961; Delone & Sandakova, 1961). However, Delone never succeeded in constructing any particular stereohedron in a space with a dimension greater than three. It was shown that a special case of a polytopic prismahedron is a gold hyper - rhombohedron of dimension 4 and higher, which is filled by a quasicrystal, revealing the periodic structure of quasicrystals in a space of higher dimension. It was shown (Zhizhin, 2017b) that a special case of a polytopic prismahedron is a gold hyper - rhombohedron of dimension 4 and higher, which is filled by a quasi - crystal, revealing the periodic structure of quasi - crystals in a space of higher dimension. By its design, a polytopic prismahedron is the product of a polytope to another polytope, or in a particular case to a one - dimensional segment. Pontryagin mentioned the structures resulting from the product of a polyhedron by a one-dimensional segment the cylinder ones (Pontryagin, 1976). We can say that the product of a polytope by a segment is a prism (Robertson, 1984) with a base in form of a polytope. To distinguish it from a usual three-dimensional prism, we call it polytopic prism. Ziegler noted that the product of polytopes is not a simplex even if the factors are simplexes, so the polytopes are of considerable interest (Ziegler, 1995). Especially taking into account that a multi-dimensional world has its own peculiarities having no analogues in the three-dimensional world (Ziegler, 1995), contrary to some opposite erroneous statements (Panina, 2006). In this regard, the theory of simplicial polytopes for the analysis of polytopes product becomes inapplicable, especially in the case of high - dimensional factors.

The product of two polytopes is result of the product of one of them by one - dimensional edges of another polytope. Thus, the product of polytopes is a complex of polytopic prism. Let`s call this complex a polytopic prismahedron. The polytopic prism is particular case of polytopic prismahedron. The interest to the study of polytopic prismahedrons is connected, in first place, with the novelty of this field and, secondly, with the fact that polytopic prismahedrons, due to their construction, can be "bricks" to fill the spaces of higher dimension face in face.

The definition of polytopes product (Ziegler, 1995) does not give the possibility to specify the structure of product of polytopes as a function of the factors structures.

THE STRUCTURE OF PRODUCT OF POLYTOPES AS FUNCTION OF FACTORS STRUCTURES

There is the structure of product of polytopes having different structures of their factors is determined.

Theorem 1 (Zhizhin, 2015)

If we have convex polytopes of dimensions *n* and *m*, respectively denoted P^n and Q^m (or simply P and Q), then their product $P^n \times Q^m$, (or simply ×, when it is clear what polytopes are multiplied) has a face $F_{\times}^k$ with numbers

$$f_{\times}^k = \sum_{i=0}^{j} f_P^{k-i} f_Q^i \qquad (1)$$

$j = k$, $j = k$, if $0 \leq k < m; j = m$, if $m \leq k \leq n + m; n \geq m$.

Here the symbol *f* indicates the number of faces, the superscript of *f* and *F* indicates the dimension of faces, the lower index indicates belonging of a face to the respective polytope.

Proof

To find the product of polytopes *P* and *Q* one have to find the product of geometric elements of different dimensions of one of the polytopes by geometric elements of different dimensions of another polytope. According to the definition of polytopes product (Ziegler, 1995), the product of vertices (elements of zero - dimension) of one of the polytopes by the vertices of another polytope is a set of vertices, the number of which is equal to the product of the number of vertices of one of the polytopes by the number of vertices of another polytope, i.e.

$$f_{\times}^0 = f_P^0 f_Q^0$$

The number of elements of dimension 1 of polytopes product consists of the product of number of elements of dimension 1 of one of polytopes by the number of elements of zero - dimension of another polytope and the product of the number of elements of zero - dimension of one of polytopes by the number of elements of the dimension 1 of another polytope, i.e.

$$f_{\times}^{1} = f_{P}^{1} f_{Q}^{0} + f_{P}^{0} f_{Q}^{1}$$

The number of elements of dimension 2 of polytopes product consists of the product of the number of elements of dimension 2 of one of the polytopes by the number of elements of zero dimension of another polytope, the product of the number of elements of dimension 1 of one of polytopes by the number of elements of dimension 1 of another polytope and the product of the number of elements of zero dimension of one of the polytopes by the number of elements of dimension 2 of another polytope, i.e.

$$f_{\times}^{2} = f_{P}^{2} f_{Q}^{0} + f_{P}^{1} f_{Q}^{1} + f_{P}^{0} f_{Q}^{2}$$

Continuing this process, we'll obtain the expression for the number of elements of dimension k in the product of polytopes P and Q

$$f_{\times}^{k} = f_{P}^{k} f_{Q}^{0} + f_{P}^{k-1} f_{Q}^{1} + \ldots + f_{P}^{0} f_{Q}^{k} \tag{2}$$

The specific form of equation (2) depends on the ratio of k with the dimensions of polytopes P and Q. If k is greater than m, i.e. the dimension of the polytope Q, then in the series (2) several last items are lost because of the absence of faces of the polytope Q with dimension greater than m. If k is greater than n, i.e. the dimension of the polytope P, in the series (2) several first items are lost because of the absence of faces of the polytope P with the dimension greater than n.

If $k < m$ the series (2) can be written in the form

$$f_{\times}^{k} = \sum_{i=0}^{k} f_{P}^{k-i} f_{Q}^{i} \tag{3}$$

When $k \geq m$ summation by i in the series (3) is carried out only up to the value $i = m$

$$f_{\times}^{k}=\sum_{i=0}^{m} f_{P}^{k-i} f_{Q}^{i} \tag{4}$$

At that, in the row (4) the items corresponding to the values f $k - i > n$ are equal to zero. Equations (3) and (4) can be written as a single equation

$$f_{\times}^{k}=\sum_{i=0}^{j} f_{P}^{k-i} f_{Q}^{i} \tag{5}$$

where $j = k,\ \ j = k$, if $0 \le k < m; j = m$, if $m \le k \le n + m; n \ge m$.

Theorem 1 is proved.

Theorem 1 determines the number of figures of different dimensions in the polytopes product, but it cannot identify in general terms what is the form of these figures. It follows from theorem 1 that the form of the figures of polytopes product depends on the form of the polytopes, which are the factors of this product.

THE PRODUCTS OF POLYTOPES BY ONE – DIMENSIONAL SEGMENT

It is obvious that the product of a point *a* by a one - dimensional segment is the same segment with point on end coinciding with point *a*. The product of segment by segment is a quadrangle. The product triangle on the segment is a triangular prism with the bases equal to the triangle, and generating lines equal to the segment.

The Product of a Tetrahedron by a Segment

For the designation of a polytope let's agree to indicate not only its dimension but the number of its vertices by lower index of polytope symbol. Besides, we'll also indicate the number of its faces in brackets after the polytope sign, face dimension and the number of its vertices (for the faces with dimension greater or equal to 2). For example, a tetrahedron in this record has the form $P_4^3(4F_3^2)$. If a polytope is considered in this case as a face we'll denote it by the letter *F*. Let's call such a record a structural polytope formula. We need to get from structural formulas of polytopes as factors the structural formula of the polytopes product. Thus, in this case, we consider the product $P_4^3(4F_3^2)\times Q_2^1$. According to theorem 1 and the ratios (5) we have

$$f_\times^0 = f_P^0 f_Q^0 = 4\cdot 2 = 8$$

$$f_\times^1 = f_P^1 f_Q^0 + f_P^0 f_Q^1 = 6\cdot 2 + 4\cdot 1 = 16.$$

When calculating the number of faces of dimension greater than 1 we'll indicate in brackets after the sign of the faces number the designation of the corresponding face. This will give us an opportunity to get the composition of faces in the polytopes product. Thus,

$$f_\times^2 = f_P^2(F_{P3}^2)f_Q^0(F_{Q1}^0) + f_P^1(F_{P2}^1)f_Q^1(F_{Q2}^1) = 4\cdot 2 + 6\cdot 1 = 14. \tag{6}$$

For receiving of the composition of the faces in the dimension 2 polytopes product we multiply the faces in brackets in the product of the numbers of faces, observing the same rules as at the definition of the polytopes product, i.e. dimensions are added and the numbers of vertices are multiplied. At that, indexes P and Q are replaced by index $\times$. Then 14 faces obtained in (6), are deciphered so

$$14 = 8F_{\times 3}^2 + 6F_{\times 4}^2. \tag{7}$$

This means that 14 faces of dimension 2 include eight faces $F_{\times 3}^2$ (i.e., triangles) and six faces $F_{\times 4}$ (i.e., quadrangles). In future, equations similar to (6), (7) will be recorded in a single line.

For the faces of dimension 3, we have

$$f_\times^3 = f_P^3(F_{P4}^3)f_Q^0(F_{Q1}^0) + f_P^2(F_{P3}^2)f_Q^1(F_{Q2}^1) = 1\cdot 2 + 4\cdot 1 = 6 = 2F_{\times 4}^3 + 4F_{\times 6}^3.$$

Thus, 6 faces of dimension 3 consist of two tetrahedrons and four triangular prisms $P_6^3(3F_4^2, 2F_3^2)$.

For the faces of dimension 4, we have

$$f_\times^4 = f_P^3(F_{P4}^3)f_Q^1(F_{Q2}^1) = 1\cdot 1 = 1 = P_8^4.$$

According to the preceding equation, we obtain the structural formula of the product

$$P_4^3(4F_3^2)\times Q_2^1 = P_8^4\left[4P_6^3\left(3F_4^2,2F_3^2\right),2F_4^3\left(4F_3^2\right)\right]. \quad (8)$$

Therefore, the product of the tetrahedron by the segment is polytope 4*D*, which includes four triangular prisms and 2 tetrahedrons. You can call a polytope (8) tetrahedral prism. The picture of the polytope (8) is shown in Figure 1.

It is easy to verify that the Euler-Pioncare equation (2) in Chapter 2 for the polytope (8) is satisfied

$8 - 16 + 14 - 6 = 0.$

In future, the verification of the Euler–Pioncare (Pioncare, 1895) equation satisfaction is omitted, as a reader can easily do it on his own.

The Product of a Cube by a Segment

The structural formula of a cube is $P_8^3(6F_4^2)$. It is necessary to find a structural formula of the product $P_8^3(6F_4^2)\times Q_2^1$. According to theorem 1 and formulas (5), we find

Figure 1. The tetrahedral prism

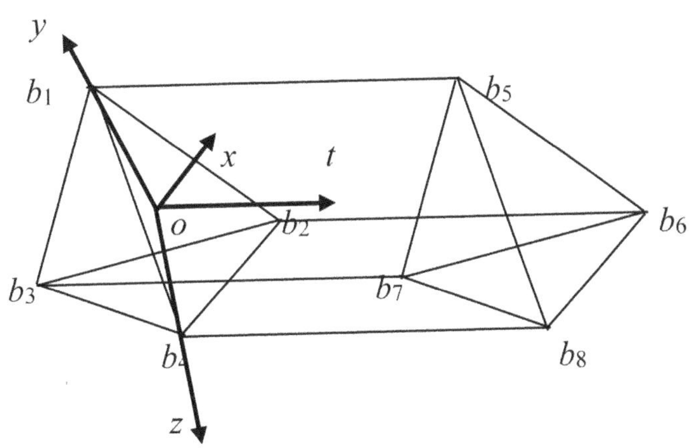

$$f_{\times}^{0} = f_P^0 f_Q^0 = 8 \cdot 2 = 16,$$

$$f_{\times}^{1} = f_P^1 f_Q^0 + f_P^0 f_Q^1 = 12 \cdot 2 + 8 \cdot 1 = 32,$$

$$f_{\times}^{2} = f_P^2(F_4^2) f_Q^0(F_{Q1}^0) + f_P^1(F_{P2}^1) f_Q^1(F_{Q2}^1) = 6 \cdot 2 + 12 \cdot 1 = 24 = 24F_{\times 4}^2,$$

$$f_{\times}^{3} = f_P^3(F_{P8}^2) f_Q^0(F_{Q1}^0) + f_P^2(F_{P4}^2) f_Q^1(F_{Q2}^1) = 1 \cdot 2 + 6 \cdot 1 = 8 = 8F_{\times 8}^3,$$

$$f_{\times}^{4} = f_P^3(F_{P8}^3) f_Q^1(F_{Q2}^1) = 1 \cdot 1 = 1 = P_{16}^4(8F_8^3). \tag{9}$$

Thus, we see that the product of a cube by a segment is $4D$ – cube (hyper - cube) containing 8 cubes.

The Product of an Octahedron by a Segment

The structural formula of an octahedron is $P_6^3(8F_4^2)$. It is necessary to find a structural formula of the product $P_6^3(8F_3^2) \times Q_2^1$. According to theorem 1 and formulas (5), we find

$$f_{\times}^{0} = f_P^0 f_Q^0 = 6 \cdot 2 = 12,$$

$$f_{\times}^{1} = f_P^1 f_Q^0 + f_P^0 f_Q^1 = 6 \cdot 1 + 12 \cdot 2 = 30,$$

$$f_{\times}^{2} = f_P^2(F_{P3}^2) f_Q^0(F_{Q1}^0) + f_P^1(F_{P2}^1) f_Q^1(F_{Q2}^1) = 8 \cdot 2 + 12 \cdot 1 = 28 = 16F_{\times 3}^2 + 12F_{\times 4}^2,$$

$$f_{\times}^{3} = f_P^3(F_{P6}^2) f_Q^0(F_{Q1}^0) + f_P^2(F_{P3}^2) f_Q^1(F_{Q2}^1) = 8F_{\times 6}^3(3F_4^2, 2F_3^2) + 2F_{\times 6}^3(8F_3^2),$$

$$f_{\times}^{4} = f_P^3(F_{P6}^3) f_Q^1(F_{Q2}^1) = 1 \cdot 1 = 1 = P_{12}^4.$$

According to the last equality we obtain the structural formula of the product

$$P_6^3(8F_3^2) \times Q_2^1 = P_{12}^4\left[8F_6^3\left(3F_4^2, 2F_3^2\right), 2F_6^3\left(8F_3^2\right)\right]. \tag{10}$$

Thus, the product of an octahedron by a segment is a polytope 4*D*, which includes 8 triangular prisms and two octahedrons. You can call the polytope (10) an octahedral prism.

The Product of an Icosahedron by a Segment

The structural formula of an icosahedron is $P^3_{12}(20F^2_3)$. It is necessary to find the structural formula of the product $P^3_{12}(20F^2_3)\times Q^1_2$. According to theorem 1 and formulas (5), we find $f^0_\times = f^0_P f^0_Q = 12 \cdot 2 = 24$,

$$f^1_\times = f^1_P f^0_Q + f^0_P f^1_Q = 30\cdot 2 + 12\cdot 1 = 72,$$

$$f^2_\times = f^2_P(F^2_{P3}) f^0_Q(F^0_{Q1}) + f^1_P(F^1_{P2}) f^1_Q(F^1_{Q2}) = 20\cdot 2 + 30\cdot 1 = 70 = 40F^2_{\times 3} + 30F^2_{\times 4},$$

$$f^3_\times = f^3_P(F^2_{P12}) f^0_Q(F^0_{Q1}) + f^2_P(F^2_{P3}) f^1_Q(F^1_{Q2}) = 1\cdot 2 + 20\cdot 1 = 22 = 20F^3_{\times 6}(3F^2_4, 2F^2_3) + 2F^3_{\times 12},$$

$$f^4_\times = f^3_P(F^3_{P12}) f^1_Q(F^1_{Q2}) = 1\cdot 1 = 1 = P^4_{24}.$$

According to the last equality, we obtain the structural formula of the product

$$P^3_{12}(20F^2_3)\times Q^1_2 = P^4_{24}\left[20F^3_6\left(3F^2_4, 2F^2_3\right), 2F^3_{12}\left(20F^2_3\right)\right]. \qquad (11)$$

Thus, the product of an icosahedron by a segment is a polytope 4*D*, which includes 20 triangular prisms and 2 icosahedrons. You can call a polytope (11) icosahedral prism.

The Product of a Dodecahedron on a Segment

The structural formula of a dodecahedron is $P^3_{20}(12F^2_5)$. It is necessary to find the structural formula of product $P^3_{20}(12F^2_5)\times Q^1_2$. According to theorem 1 and formulas (5), we find

$$f^0_\times = f^0_P f^0_Q = 20 \cdot 2 = 40,$$

$$f_{\times}^{1} = f_{P}^{1} f_{Q}^{0} + f_{P}^{0} f_{Q}^{1} = 30 \cdot 2 + 20 \cdot 1 = 80,$$

$$f_{\times}^{2} = f_{P}^{2}(F_{P5}^{2}) f_{Q}^{0}(F_{Q1}^{0}) + f_{P}^{1}(F_{P2}^{1}) f_{Q}^{1}(F_{Q2}^{1}) = 12 \cdot 2 + 30 \cdot 1 = 54 = 24F_{\times 5}^{2} + 30F_{\times 4}^{2},$$

$$f_{\times}^{3} = f_{P}^{3}(F_{P20}^{2}) f_{Q}^{0}(F_{Q1}^{0}) + f_{P}^{2}(F_{P5}^{2}) f_{Q}^{1}(F_{Q2}^{1}) = 1 \cdot 2 + 12 \cdot 1 = 14 = 12F_{\times 10}^{3}(5F_{4}^{2}, 2F_{5}^{2}) + 2F_{\times 20}^{3}(12F_{5}^{2}),$$

$$f_{\times}^{4} = f_{P}^{3}(F_{P20}^{3}) f_{Q}^{1}(F_{Q2}^{1}) = 1 \cdot 1 = 1 = P_{40}^{4}.$$

According to the last equality, we obtain the structural formula of the product

$$P_{20}^{3}(12F_{5}^{2}) \times Q_{2}^{1} = P_{40}^{4}\left[12F_{10}^{3}\left(5F_{4}^{2}, 2F_{5}^{2}\right), 2F_{20}^{3}\left(12F_{5}^{2}\right)\right]. \tag{12}$$

Thus, the product of a dodecahedron by a segment is a polytope $4D$ including 12 pentagonal prisms and 2 dodecahedrons. You can call the polytope (12) a dodecahedral prism.

The Product of an Egyptian Pyramid by a Segment

The structural formula of an Egyptian pyramid is $P_{5}^{3}\left(4F_{3}^{2}, \mathrm{F}_{4}^{2}\right)$. It is necessary to find the structural formula of product $P_{5}^{3}\left(4F_{3}^{2}, \mathrm{F}_{4}^{2}\right) \times Q_{2}^{1}$. According to theorem 1 and formulas (5), we find

$$f_{\times}^{0} = f_{P}^{0} f_{Q}^{0} = 5 \cdot 2 = 10,$$

$$f_{\times}^{1} = f_{P}^{1} f_{Q}^{0} + f_{P}^{0} f_{Q}^{1} = 5 \cdot 1 + 8 \cdot 2 = 21,$$

$$f_{\times}^{2} = f_{P}^{2}(4F_{P3}^{2}, \mathrm{F}_{P4}^{2}) f_{Q}^{0}(F_{Q1}^{0}) + f_{P}^{1}(F_{P2}^{1}) f_{Q}^{1}(F_{Q2}^{1}) = 5 \cdot 2 + 8 \cdot 1 = 18 = 8F_{\times 3}^{2} + 10F_{\times 4}^{2},$$

$$f_{\times}^{3} = f_{P}^{3}(F_{P5}^{2}) f_{Q}^{0}(F_{Q1}^{0}) + f_{P}^{2}(F_{P3}^{2}, \mathrm{F}_{P4}^{2}) f_{Q}^{1}(F_{Q2}^{1}) = 1 \cdot 2 + 5 \cdot 1 = 7 = 2F_{\times 5}^{3} + 4F_{\times 6}^{2} + F_{\times 8}^{3},$$

$$f_\times^4 = f_P^3(F_{P5}^3)f_Q^1(F_{Q2}^1) = 1 \cdot 1 = 1 = P_{10}^4.$$

According to the last equality, we obtain the structural formula of the product

$$P_5^3(4F_3^2, \mathrm{F}_4^2) \times Q_2^1 = P_{10}^4\left[4F_6^3\left(3F_4^2, 2F_3^2\right), F_8^3\left(6F_4^2\right), 2\,\mathrm{F}_5^3\left(4\,\mathrm{F}_3^2, \mathrm{F}_4^2\right)\right].$$

Thus, the product of an Egyptian pyramid by segment is a polytope 4*D* including four triangular prisms, 1 quadrangular prism and 2 Egyptian pyramids (Figure 2). You can call the polytope a pyramidal prism.

It is interesting that the pyramids of ancient Egypt had a strict hemetric form (the pyramids of Cheops, Khafre, Mikerin) and were built in groups not far from each other (Snegirev & Franzev, 1938). In aggregate, this led, according to the analysis carried out, to an increase in the dimensionality of the space in the vicinity of the pyramids.

The Product of a 4 - Simplex by a Segment

The structural formula of a 4-simplex is $P_5^3\left[5F_4^3\left(4F_3^2\right)\right]$. It is necessary to find the structural formula of the product $P_5^3\left[5F_4^2\left(4F_3^2\right)\right] \times Q_2^1$. According to theorem 1 and formulas (5), we find $f_\times^0 = f_P^0 f_Q^0 = 5 \cdot 2 = 10$,

$$f_\times^1 = f_P^1 f_Q^0 + f_P^0 f_Q^1 = 10 \cdot 2 + 5 \cdot 1 = 25,$$

$$f_\times^2 = f_P^2(F_{P3}^2)f_Q^0(F_{Q1}^0) + f_P^1(F_{P2}^1)f_Q^1(F_{Q2}^1) = 10 \cdot 2 + 10 \cdot 1 = 30 = 20F_{\times 3}^2 + 10F_{\times 4}^2,$$

$$f_\times^3 = f_P^3(F_{P4}^2)f_Q^0(F_{Q1}^0) + f_P^2(F_{P3}^2)f_Q^1(F_{Q2}^1) = 5 \cdot 2 + 10 \cdot 1 = 20 = 10F_{\times 4}^3(4F_3^2) + 10F_{\times 6}^2(3F_4^2, 2F_3^2),$$

$$f_\times^4 = f_P^4[5F_{P4}^3(4F_3^3)]f_Q^0(F_{Q1}^0) + f_P^2(F_{P4}^2)f_Q^1(F_{Q2}^1) = 1 \cdot 2 + 5 \cdot 1 = 7 =$$
$$5F_{\times 8}^4[4F_6^3(3F_4^2, 2F_3^2), 2F_4^3(4F_3^2)] + 2F_{\times 5}^2[5F_4^3(4F_3^2)],$$

$$f_\times^5 = f_P^4(F_{P5}^3)f_Q^1(F_{Q2}^1) = 1 \cdot 1 = 1 = P_{10}^5.$$

Figure 2. The pyramidal prism

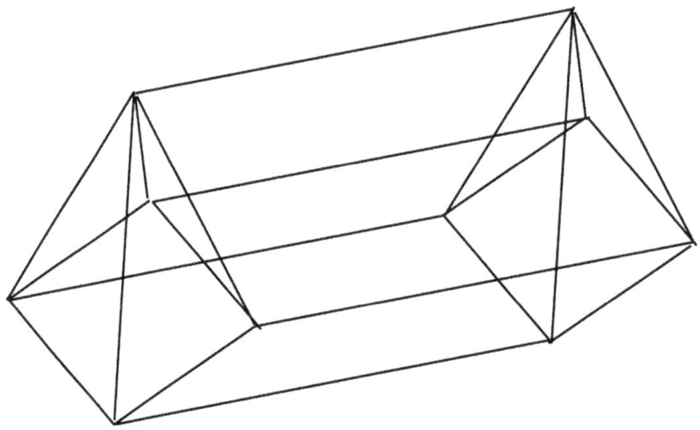

According to the last equality, we obtain the structural formula of the product

$$P_5^3[5F_4^3(4F_3^2)]\times Q_2^1 = P_{10}^4\{5F_8^4[4F_6^3(3F_4^2,2F_3^2),2F_4^3(4F_3^2)],2F_5^4[5F_4^3(4F_3^2)]\}. \tag{13}$$

Thus, the product of a 4 - simplex by a segment is a polytope 5*D* including 5 tetrahedral prisms (8) and 2 4 - simplexes. We can call the polytope (13) a 5 – simplex - prism. The image of 5 –simplex - prism is shown in Figure 3.

Figure 3. The 5 – simplex - prism

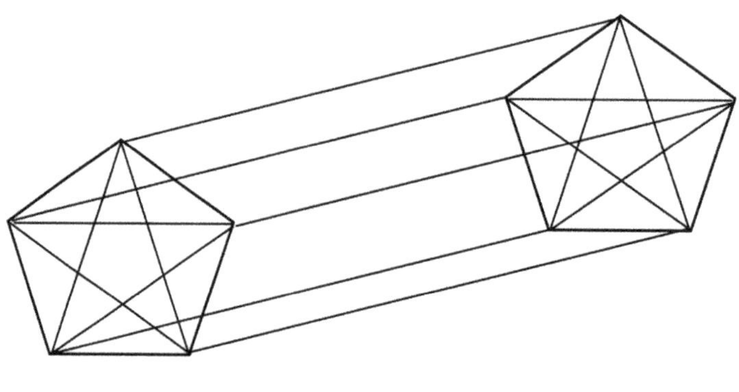

The Product of a 4 – Cross - Polytope by a Segment

The structural formula of a 4 – cross - polytope is $P_8^3[16F_4^3(4F_3^2)]$. It is necessary to find the structural formula of the product $P_8^3[16F_4^2(4F_3^2)]\times Q_2^1$. According to theorem 1 and formulas (5), we find

$$f_\times^0 = f_P^0 f_Q^0 = 8\cdot 2 = 16,$$

$$f_\times^1 = f_P^1 f_Q^0 + f_P^0 f_Q^1 = 24\cdot 2 + 8\cdot 1 = 56,$$

$$f_\times^2 = f_P^2(F_{P3}^2)f_Q^0(F_{Q1}^0) + f_P^1(F_{P2}^1)f_Q^1(F_{Q2}^1) = 32\cdot 2 + 24\cdot 1 = 88 = 64F_{\times 6}^2 + 24F_{\times 4}^2,$$

$$f_\times^3 = f_P^3(F_{P4}^2)f_Q^0(F_{Q1}^0) + f_P^2(F_{P3}^2)f_Q^1(F_{Q2}^1) = 16\cdot 2 + 32\cdot 1 = 64 = 32F_{\times 4}^3(4F_3^2) + 32F_{\times 6}^2(3F_4^2, 2F_3^2),$$

$$f_\times^4 = f_P^4\{\mathrm{F}_8^4[16F_{P4}^3(4F_3^3)]\}f_Q^0(F_{Q1}^0) + f_P^2(F_{P4}^2)f_Q^1(F_{Q2}^1) = 1\cdot 2 + 16\cdot 1 = 18 =$$
$$16F_{\times 8}^4[4F_6^3(3F_4^2, 2F_3^2), 2F_4^3(4F_3^2)] + 2F_{\times 8}^2[16F_4^3(4F_3^2)],$$

$$f_\times^5 = f_P^4\{\mathrm{F}_4^3[16F_4^3(4F_3^3)f_Q^1(F_{Q2}^1) = 1\cdot 1 = 1 = P_{16}^5.$$

According to the last equality, we obtain the structural formula of the product

$$P_5^3[5F_8^3[16F_4^3(4F_3^2)]\times Q_2^1 = P_{16}^4\{16F_8^4[4F_6^3(3F_4^2, 2F_3^2), 2F_4^3(4F_3^2)], 2F_8^4[16F_4^3(4F_3^2)]\}. \tag{14}$$

Thus, the product of a 4 – cross - polytope by a segment is a polytope 5*D* including 16 tetrahedral prisms (8) and 2 4 – cross - polytopes. We can call the polytope (14) a 5 – cross - prism (Figure 4).

The General Structural Formula of Polytopes Product by a Segment

We'll call prisms with bases in the form of polytopes the polytopic prisms.

Figure 4. The 5 - cross - prism

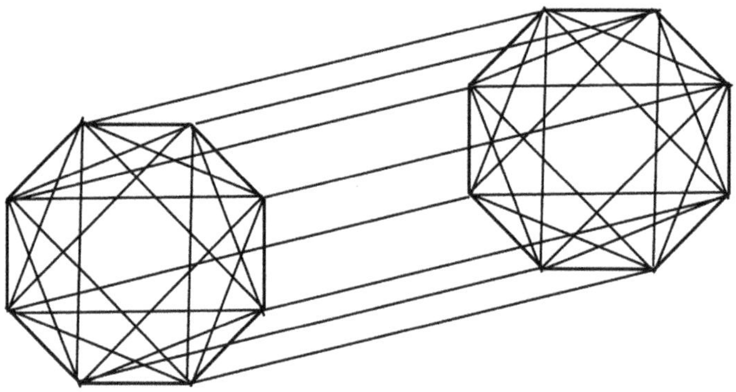

Theorem 2

The product of a polytope by a segment, if all its faces of dimension 2 are equal, is a polytopic prism with the structure defined by equation

$$\begin{aligned}&P_{K_n}^{n}\{f_{n-1}F_{K_{n-1}}^{n-1}...[\mathrm{f}_3\, F_{K_3}^{3}(f_2F_{K_2}^{2})]...\}\times Q_2^1=\\&P_{2K_n}^{n+1}\{f_{n-1}F_{2K_{n-1}}^{n}...[f_3F_{2K_3}^{4}(f_2F_{2K_2}^{3},2F_{K_2}^{2})2F_{K_3}^{3}]...2F_{K_n}^{n}\}.\end{aligned} \quad (15)$$

There $P_{K_n}^{n}$ is the polytope of dimension *n* with the number of vertices K_n; $F_{K_{n-1}}^{n-1}$ are hyper-faces with the number f_{n-1}, each having K_{n-1} vertices; accordingly $F_{K_2}^{2}$ are hyper - faces of dimension 2 at faces $F_{K_3}^{3}$; f_2 is the number of faces $F_{K_2}^{2}$ with the number of vertices K_2 and so on; $P_{2K_n}^{n+1}$ is a polytopic prism of dimension *n+1* with the number of vertices $2\,K_n$, the bases of which are the polytopes $P_{K_n}^{n}$. The polytopic prism $P_{2K_n}^{n+1}$ includes f_{n-1} polytopic prisms of dimension *n* and the number of the vertices $2\,K_{n-1}$, and also two polytopes $P_{K_n}^{n}$ which are the bases of polytopic prisms and so on, up to prisms $F_{2K_2}^{3}$ with a number f_2 and with bases $F_{K_2}^{2}$.

Proof

Theorem 1 follows from the carried out examinations of the product of polytopes of different dimensions by a segment. General laws which allowed us to formulate the theorem can be traced in all these products. Whatever the dimension *n* of the polytope was, at multiplying it by a segment of each vertex of the polytope, the segment which is multiplied by a polytope as if grows. This forms polytopic prism, the bases of which are two copies of the initial polytope. The dimension of this polytopic prism is by 1 more than the dimension of the initial polytope, i.e. $n + 1$. This polytopic prism includes polytopic prisms built on the hyper-faces of the initial polytope and their number is equal to the number of hyper-faces of the initial polytope f_{n-1}. The dimension of these politopic prisms is *n*, since the dimension of the initial polytope hyper-faces is equal to *n* - 1, but when multiplied by a segment the dimension is increased by 1. In addition, a polytopic prism of dimension *n* + 1 consists of two copies of the initial polytope of dimension *n*, as the bases of the prism. In its turn, the polytopic prisms of dimension *n* include the polytopic prisms of dimension *n* -1, built on the hyper-faces of the polytope of dimension *n* - 1, and two copies of the polytope of dimension *n* - 1. This process of the polytopic prisms inclusion in each other happens until the formation of three - dimensional prisms built on two - dimensional facets of the initial polytope.

The equality (15) is an analytical expression of these constructions. This process of formation occurs when all two - dimensional faces of the initial polytope are the same, i.e. they are equal polygons. If the initial polytope has two - dimensional faces with different numbers of vertices so the analytical record of these polytopes product by a segment will be added by prisms built on different bases. The ratios (15) don't include such options of polytopic prisms.

Theorem 2 it is proved.

Supplement. Values of f_i in (15) depend on the form and dimension of the polytope. According to Zhizhin (2014) the formulas for their calculation in the case of a simplex, cube and cross - polytope will be the following: for simplex $f_i = C_{i+1}^i = i + 1$; for cube $f_i = 2C_i^{i-1} = 2i$; for cross - polytope $f_i = 2^i, i > 3, f_3 = 4$.

Naturally, the results of research about products of polytopes by a segment are to be taken into account when analyzing the products of polytopes by more complex figures, since they all have edges are factors in these products.

BOUNDARY COMPLEXES AND INTERIOR POINTS OF POLYTOPIC PRISMS

Using the results of Chapter 2, we consider in more detail the geometric properties of polytopic prisms. Let's turn to the simplest of polytopic prisms - the tetrahedral prism (Figure 1) and decide whether the tetrahedral prism has internal points, i.e. points not belonging to its boundary complex. For this purpose, in Figure 1, coordinate systems are introduced, as well as in Figure 6 of Chapter 2, when the tetrahedron is considered. Taking into account that in this case we are dealing with a space of dimension 4, a fourth coordinate *t* is added to the coordinate system parallel to the generating lines of the tetrahedral prism. Assuming the length of all edges of the tetrahedral prism per unit, we can establish the coordinates of all the vertices of the tetrahedral prism (Figure 1) in the given coordinate system (*x, y, z, t*)

$$b_1\left(0,\frac{1}{\sqrt{3}},0,0\right),b_2\left(\frac{1}{2},-\frac{1}{2\sqrt{3}},0,0\right),b_3\left(-\frac{1}{2},-\frac{1}{2\sqrt{3}},0,0\right),b_4\left(0,0,\sqrt{\frac{2}{3}},0\right),$$

$$b_5\left(0,\frac{1}{\sqrt{3}},0,1\right),b_6\left(\frac{1}{2},-\frac{1}{2\sqrt{3}},0,1\right),b_7\left(-\frac{1}{2},-\frac{1}{2\sqrt{3}},0,1\right),b_8\left(0,0,\sqrt{\frac{2}{3}},1\right).$$

Then the set of interior points of the tetrahedral prism is determined by the system of inequalities (see Chapter 2)

$$z>0,0<t<1;$$

$$z<2\sqrt{\frac{2}{3}}x-\sqrt{2}y+\sqrt{\frac{2}{3}},-\frac{1}{2}<x<0,-\frac{1}{2\sqrt{3}}<y<\frac{1}{\sqrt{3}},0<\mathrm{t}<1;$$

$$z<2\sqrt{2}y+\sqrt{\frac{2}{3}},-\frac{1}{2}<x<\frac{1}{2};-\frac{1}{2\sqrt{3}}<y<0,0<t<1;$$

$$z < -\sqrt{6}x - \sqrt{2}y + \sqrt{\frac{2}{3}}, 0 < x < \frac{1}{2}, -\frac{1}{2\sqrt{3}} < y < \frac{1}{\sqrt{3}}, 0 < t < 1.$$

Similarly, in principle, you can define a set of internal points and other polytopic prisms.

Following Chapter 2 from Figure 1 of this Chapter 3, we find the incidence of the elements of a boundary complex of different dimensions of a tetrahedral prism

$$k_{b(i=0)}(4) = 4, k_{b(i=1)}(4) = 3, k_{b(i=2)}(4) = 2, k_{b(i=3)}(4) = 1.$$

Thus, the incidence values of the elements of the boundary complex of the tetrahedral prism obey the formula (3) and Theorem 2 of Chapter 2.

The same incidence values have elements of the boundary complex of the pyromidal prism (Figure 2) with one exception: the tops of the Egyptian pyramids have an incidence value

$$k_{b(i=0)}(4) = 5.$$

Polytopic prisms with increasing dimension, if they do not include cross - polytopes, also obey Theorem 2 and formula (5) of Chapter 2. For example, the 5 – simplex - prism (Figure 3) has the following incidence of elements of its boundary complex

$$k_{b(i=0)}(5) = 5, k_{b(i=1)}(5) = 4, k_{b(i=2)}(5) = 3, k_{b(i=3)}(5) = 2, k_{b(i=4)}(5) = 1.$$

This agrees with formula (3) and Theorem 2 of Chapter 2.

However, if the cross - polytope is included in the boundary polytope prism complex, the situation changes. For example, Figure 4 shows an image of a 5 – cross - prism, which includes two 4 – cross - polytopes. From this image it follows that in this case the incidence of the elements of the boundary complex have the values

$$k_{b(i=0)}(5) = 7, k_{b(i=1)}(5) = \begin{cases} 6 \\ 5 \end{cases}, k_{b(i=2)}(5) = 4, k_{b(i=3)}(5) = 2, k_{b(i=4)}(5) = 1.$$

The maximum value of incidence of edge 6 refers to the forming lines of the 5 – cross - prism, the minimum incidence value of edge 4 refers to the edges of 4 – cross - polytopes.

THE PRODUCTS OF POLYTOPES BY A TRIANGLE

The Product of a Triangle by a Triangle

There we consider the product $P_3^2 \times Q_3^2$. According to theorem 1 and ratios (5) we have

$$f_\times^0 = f_P^0 f_Q^0 = 3 \cdot 3 = 9,$$

$$f_\times^1 = f_P^1 f_Q^0 + f_P^0 f_Q^1 = 3 \cdot 3 + 3 \cdot 3 = 18,$$

$$f_\times^2 = f_P^2(4F_{P3}^2) f_Q^0(F_{Q1}^0) + f_P^1(F_{P2}^1) f_Q^1(F_{Q2}^1) + f_P^0(F_{P1}^0) f_Q^2(F_{Q3}^2) = 1 \cdot 3 + 3 \cdot 3 + 3 \cdot 1 = 15 = 6F_{\times 3}^2 + 9F_{\times 4}^2,$$

$$f_\times^3 = f_P^3(F_{P3}^2) f_Q^1(F_{Q2}^1) + f_P^1(F_{P2}^1) f_Q^2(F_{Q3}^2) = 1 \cdot 3 + 3 \cdot 1 = 6 = 6F_{\times 6}^3,$$

$$f_\times^4 = f_P^2(F_{P3}^2) f_Q^2(F_{Q3}^2) = 1 \cdot 1 = 1 = P_{\times 9}^4.$$

Thus, the product of a triangle by a triangle is 4 - polytope having 9 vertices and 6 triangle prisms. The existence of such polytope was proved of Ziegler (1995) independently.

In particular, the fundamental difference between this polytope and simplicial polytopes and between this polytope and three - dimensional polyhedrons because of its multidimensionality is indicated of Ziegler (1995). The structural formula of this polytope (the triangular prismahedron) has the form

$$P_3^2(F_{P3}^2) \times Q_3^2(F_{Q3}^2) = P_9^4[6P_6^3(3F_4^2, F_3^2)]. \tag{16}$$

The image of a triangular prismahedron (Zhizhin, 2017) is shown in Figure 5.

Figure 5. The triangular prismahedron

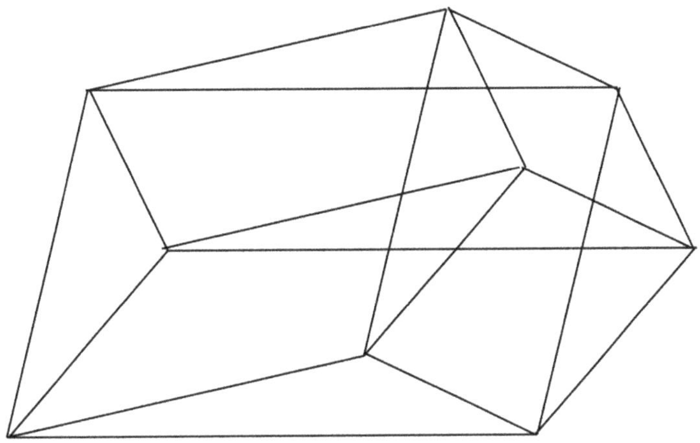

The Product of a Polygon by a Triangle

According to theorem 1 and ratios (5) for the product $P_i^2(F_{Pi}^2)\times Q_3^2(F_{Q3}^2)$ (i is the number of the polygon sides) we have

$$f_\times^0 = f_P^0 f_Q^0 = i\cdot 3 = 3i\,,$$

$$f_\times^1 = f_P^1 f_Q^0 + f_P^0 f_Q^1 = i\cdot 3 + 3\cdot i = 6i,$$

$$f_\times^2 = f_P^2(F_{Pi}^2)f_Q^0(F_{Q1}^0) + f_P^1(F_{P2}^1)f_Q^1(F_{Q2}^1) + f_P^0(F_{P1}^0)f_Q^2(F_{Q3}^2) = 1\cdot 3 + i\cdot 3 + i\cdot 1 = 3 + 4i = 3F_{\times i}^2 + 3iF_{\times 4}^2 + iF_{\times 3}^2,$$

$$f_\times^3 = f_P^3(F_{Pi}^2)f_Q^1(F_{Q2}^1) + f_P^1(F_{P2}^1)f_Q^2(F_{Q3}^2) = 1\cdot 3 + i\cdot 1 = 3 + i = iF_{\times 6}^3 + 3F_{\times 2i}^3,$$

$$f_\times^4 = f_P^2(F_{Pi}^2)f_Q^2(F_{Q3}^2) = 1\cdot 1 = 1 = P_{\times 3i}^4.$$

Thus, the product of the polygon F_i^2 by the triangle F_3^2 is a 4 - polytope having $3i$ vertices, 3 prisms with bases F_i^2 and i prisms with triangle bases. The structural formula of this product has the form

$$P_i^3(F_{Pi}^2)\times Q_3^2(F_{Q3}^2)=P_{3i}^4[3F_{2i}^3(\mathrm{i}\,F_4^2,2F_3^2),\mathrm{i}\,F_6^3(3F_4^2,2F_3^2)]. \quad (17)$$

Examples of products of a square, a pentagon and a hexagon by a triangle are shown in Figures 6 – 8.

The polytope (17) is a generalization of the polytope (16) in the case when one of the multiplied polygons has the number of sides greater than 3. This polytope in the general case can be called 3 *i - angular prismahedron. Therefore, the polytope (16) can be more accurately described as 3*3 - angular prismahedron.

The Product of a Tetrahedron by a Triangle

According to theorem 1 and ratios (5) for the numbers of elements of different dimension in the product $P_4^2(4F_3^2)\times Q_3^2$ we have

$$f_\times^0=f_P^0 f_Q^0=4\cdot 3=12,$$

$$f_\times^1=f_P^1 f_Q^0+f_P^0 f_Q^1=6\cdot 3+3\cdot 4=30,$$

*Figure 6. The 4*3 – angular prismahedron*

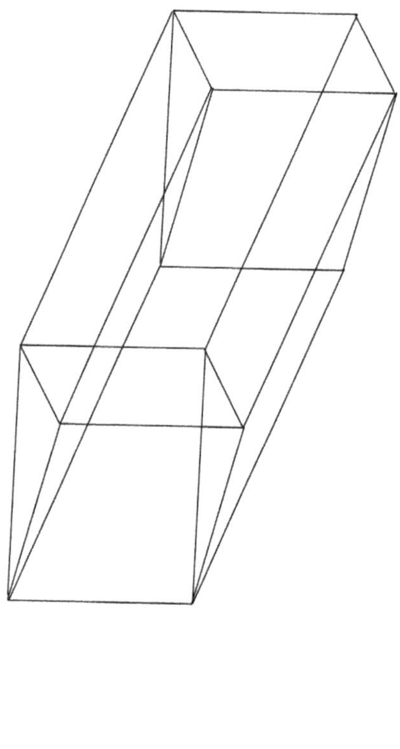

*Figure 7. The 5*3 – angular prismahedron*

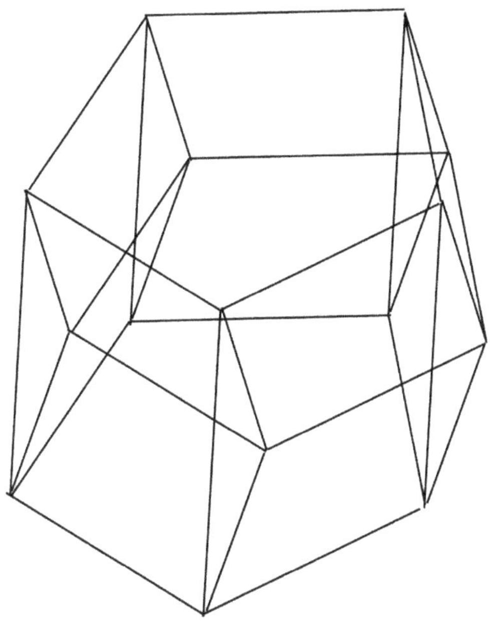

*Figure 8. The 6*3 – angular prismahedron*

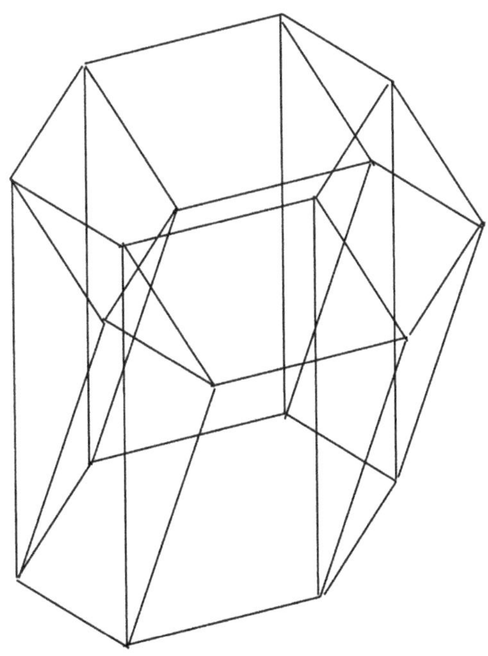

$$f_\times^2 = f_P^2(F_{P3}^2)f_Q^0(F_{Q1}^0) + f_P^1(F_{P2}^1)f_Q^1(F_{Q2}^1) + f_P^0(F_{P1}^0)f_Q^2(F_{Q3}^2) = 4\cdot 1 + 6\cdot 3 + 4\cdot 1 = 26 = 8F_{\times 3}^2 + 18F_{\times 4}^2,$$

$$f_\times^3 = f_P^3(F_{P4}^2)f_Q^0(F_{Q1}^0) + f_P^2(F_{P3}^2)f_Q^1(F_{Q2}^1) + f_P^1(F_{P2}^1)f_Q^2(F_{Q3}^2) = 1\cdot 3 + 4\cdot 1 + 6\cdot 1 = 21 = 18F_{\times 6}^3 + 3F_{\times 4}^3,$$

$$f_\times^4 = f_P^3(F_{P4}^3)f_Q^1(F_{Q2}^1) + f_P^2(F_{P3}^2)f_Q^2(F_{Q3}^2) = 1\cdot 3 + 4\cdot 1 = 7 = 3F_{\times 8}^4 + 4F_{\times 9}^4,$$
$$f_\times^5 = f_P^3(F_{P4}^3)f_Q^2(F_{Q3}^2) = 1\cdot 1 = 1 = P_{\times 12}^5.$$

Thus, the product of a tetrahedron by a triangle is a 5-polytope consisting from 3 tetrahedron prisms (8) and 4 triangular prismahedrons (16). The structural formula of this polytope has the form

$$P_4^3(4F_3^2)]\times Q_3^2 = P_{12}^5\{4F_9^4[6F_6^3(3F_4^2,2F_3^2),2F_8^4[4F_6^3(3F_4^2,2F_3^2),2F_4^3(4F_3^2)]\}. \tag{18}$$

We can call the polytope (18) a tetrahedral prismahedron. The image of this polytope on a two-dimension plane is shown in Figure 9.

The Product of an Octahedron by a Triangle

According to theorem 1 and ratios (5) for the numbers of elements of different dimension in the product $P_6^2(8F_3^2)\times Q_3^2$ we have

$$f_\times^0 = f_P^0 f_Q^0 = 6\cdot 3 = 18,$$

$$f_\times^1 = f_P^1 f_Q^0 + f_P^0 f_Q^1 = 12\cdot 3 + 6\cdot 3 = 54,$$

$$f_\times^2 = f_P^2(F_{P3}^2)f_Q^0(F_{Q1}^0) + f_P^1(F_{P2}^1)f_Q^1(F_{Q2}^1) + f_P^0(F_{P1}^0)f_Q^2(F_{Q3}^2) = 8\cdot 1 + 12\cdot 3 + 6\cdot 1 = 66 = 30F_{\times 3}^2 + 36F_{\times 4}^2,$$

$$f_\times^3 = f_P^3(F_{P6}^2)f_Q^0(F_{Q1}^0) + f_P^2(F_{P3}^2)f_Q^1(F_{Q2}^1) + f_P^1(F_{P2}^1)f_Q^2(F_{Q3}^2) = 1\cdot 3 + 8\cdot 3 + 12\cdot 1 = 39$$
$$= 36F_{\times 6}^3(3F_4^2,2F_3^2) + 3F_{\times 6}^3(8F_3^2),$$

Figure 9. The tetrahedral prismahedron

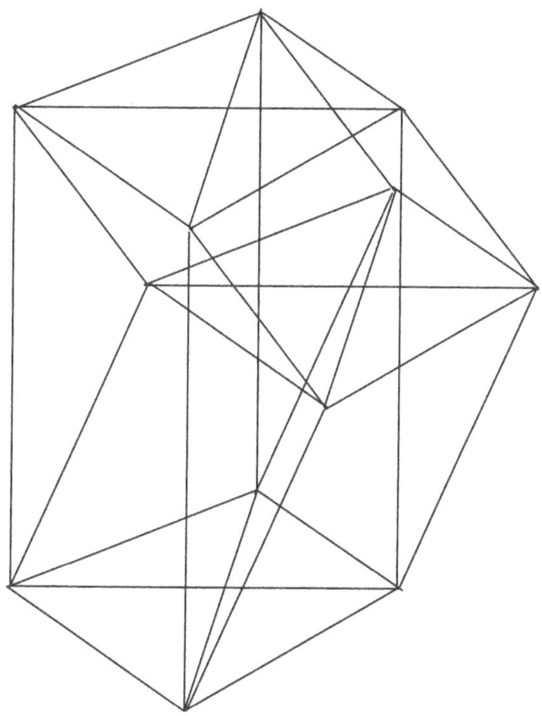

$$f_\times^4 = f_P^3[F_{P6}^3(8F_3^2)]f_Q^1(F_{Q2}^1) + f_P^2(F_{P3}^2)f_Q^2(F_{Q3}^2) = 1\cdot 3 + 8\cdot 1 = 11$$
$$= 3F_{\times 12}^4[8F_6^3(3F_4^2, 2F_3^2), 2F_6^3(8F_3^2)] + 8F_{\times 9}^4[6F_6^3(3F_4^2, 2F_3^2)],$$
$$f_\times^5 = f_P^3[F_{P6}^3(8F_3^2)]f_Q^2(F_{Q3}^2) = 1\cdot 1 = 1 = P_{\times 18}^5.$$

Thus, the product of an octahedron by a triangle is a 5 - polytope including 8 3 × 3 - angular prismahedrons (16) and 3 octahedral prisms (10).

The structural formula of the polytope P_{18}^5 has the form

$$P_6^3(8F_3^2)] \times Q_3^2 = P_{18}^5 \left\{8F_9^4[6F_6^3(3F_4^2, 2F_3^2), 3F_{12}^3\left[8F_6^3(3F_4^2, 2F_3^2), 2F_6^3(8F_3^2)\right]\right\}. \tag{19}$$

We can call the polytope (19) an octahedral prismahedron. The image of the polytope P_{18}^5 is shown in Figure 10.

Figure 10. The octahedral prismahedron

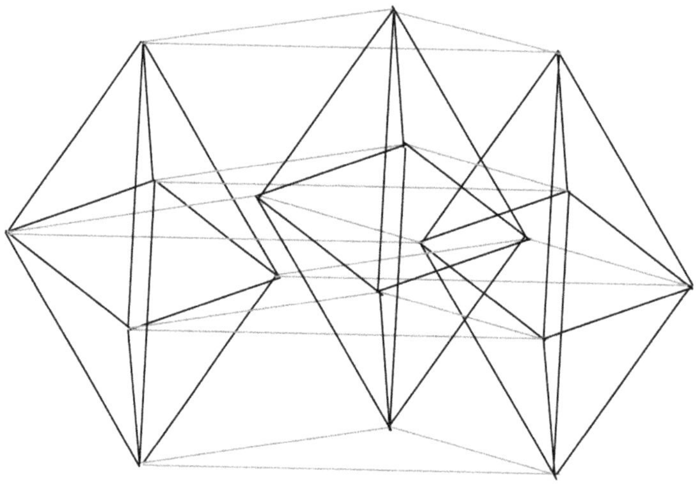

The Product of a Cube by a Triangle

According to theorem 1 and ratios (5) for the numbers of elements of different dimension in the product $P_8^2(6F_4^2) \times Q_3^2$ we have

$$f_\times^0 = f_P^0 f_Q^0 = 8 \cdot 3 = 24\,,$$

$$f_\times^1 = f_P^1 f_Q^0 + f_P^0 f_Q^1 = 12 \cdot 3 + 8 \cdot 3 = 60,$$

$$f_\times^2 = f_P^2(F_{P4}^2) f_Q^0(F_{Q1}^0) + f_P^1(F_{P2}^1) f_Q^1(F_{Q2}^1) + f_P^0(F_{P1}^0) f_Q^2(F_{Q3}^2) = 6 \cdot 3 + 12 \cdot 3 + 8 \cdot 1 = 62 = 8F_{\times 3}^2 + 54F_{\times 4}^2,$$

$$f_\times^3 = f_P^3(F_{P8}^2) f_Q^0(F_{Q1}^0) + f_P^2(F_{P4}^2) f_Q^1(F_{Q2}^1) + f_P^1(F_{P2}^1) f_Q^2(F_{Q3}^2) = 1 \cdot 3 + 6 \cdot 3 + 12 \cdot 1 = 33$$
$$= 21F_{\times 8}^3 + 12F_{\times 6}^3,$$

$$f_\times^4 = f_P^3(F_{P8}^3) f_Q^1(F_{Q2}^1) + f_P^2(F_{P3}^2) f_Q^2(F_{Q3}^2) = 1 \cdot 3 + 6 \cdot 1 = 9$$
$$= 3F_{16}^4 + 6F_{12}^4,$$

$$f_\times^5 = f_P^3(F_{P8}^3) f_Q^2(F_{Q3}^2) = 1 \cdot 1 = 1 = P_{\times 24}^5.$$

Thus, the product of a cube by a triangle is a 5 - polytope including 3 hyper - cubes (9) and 6 3 × 4 - angular prismahedrons (17).

The structural formula of the polytope P_{24}^{5} has the form

$$P_{24}^{5}\{3F_{16}^{4}(80F_{4}^{2}),6F_{12}^{4}[3F_{8}^{3}(4F_{4}^{2},2F_{3}^{2}),4F_{6}^{3}(3F_{4}^{2},2F_{3}^{2})]\}. \quad (20)$$

We can call the polytope (20) a cube - polytopic prismahedron. The image of the polytope P_{24}^{5} is shown in Figure 11.

The Product of a Dodecahedral by a Triangle

According to theorem 1 and ratios (5) for the numbers of elements of different dimension in the product $P_{20}^{3}(12F_{5}^{2})\times Q_{3}^{2}$ we have

$$f_{\times}^{0}=f_{P}^{0}f_{Q}^{0}=20\cdot 3=60,$$

$$f_{\times}^{1}=f_{P}^{1}f_{Q}^{0}+f_{P}^{0}f_{Q}^{1}=30\cdot 3+20\cdot 3=150,$$

$$f_{\times}^{2}=f_{P}^{2}(F_{P5}^{2})f_{Q}^{0}(F_{Q1}^{0})+f_{P}^{1}(F_{P2}^{1})f_{Q}^{1}(F_{Q2}^{1})+f_{P}^{0}(F_{P1}^{0})f_{Q}^{2}(F_{Q3}^{2})=12\cdot 3+30\cdot 3+20\cdot 1=146$$
$$=36F_{\times 5}^{2}+90F_{\times 4}^{2}+20F_{\times 3}^{2},$$

Figure 11. The cube - polytopic prismahedron

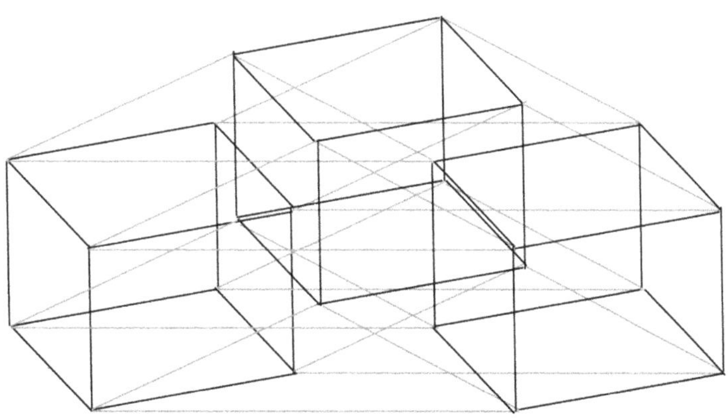

$$f_{\times}^{3} = f_{P}^{3}(F_{P20}^{3})f_{Q}^{0}(F_{Q1}^{0}) + f_{P}^{2}(F_{P5}^{2})f_{Q}^{1}(F_{Q2}^{1}) + f_{P}^{1}(F_{P2}^{1})f_{Q}^{2}(F_{Q3}^{2}) = 1\cdot 3 + 12\cdot 3 + 30\cdot 1 = 69$$
$$= 3F_{\times 20}^{3} + 36F_{\times 10}^{3} + 30F_{\times 6}^{3},$$

$$f_{\times}^{4} = f_{P}^{3}(F_{P20}^{3})f_{Q}^{1}(F_{Q2}^{1}) + f_{P}^{2}(F_{P5}^{2})f_{Q}^{2}(F_{Q3}^{2}) = 1\cdot 3 + 12\cdot 1 = 15$$
$$= 3F_{\times 40}^{4} + 12F_{\times 15}^{4},$$

$$f_{\times}^{5} = f_{P}^{3}(F_{P20}^{3})f_{Q}^{2}(F_{Q3}^{2}) = 1\cdot 1 = 1 = P_{\times 60}^{5}.$$

Thus, the product of a dodecahedron by a triangle is a 5 - polytope including 3 dodecahedral prisms (12) and 12 3*5 - angular prismahedrons (17).

The structural formula of the polytope P_{60}^{5} has the form

$$P_{60}^{5}\{3F_{40}^{4}[12F_{10}^{3}(5F_{4}^{2},2F_{3}^{2}),2F_{20}^{3}(12F_{5}^{2})],12F_{15}^{4}[3F_{10}^{3}(5F_{4}^{2},2F_{5}^{2}),5F_{6}^{3}(3F_{4}^{2},2F_{3}^{2})]\}. \tag{21}$$

We can call the polytope (21) a dodeca-polytopic prismahedron.

The Product of an Icosahedron by a Triangle

According to theorem 1 and ratios (5) for the numbers of elements of different dimension in the product $P_{12}^{3}(20F_{5}^{2})\times Q_{3}^{2}$ we have

$$f_{\times}^{0} = f_{P}^{0}f_{Q}^{0} = 12\cdot 3 = 36,$$

$$f_{\times}^{1} = f_{P}^{1}f_{Q}^{0} + f_{P}^{0}f_{Q}^{1} = 30\cdot 3 + 12\cdot 3 = 126,$$

$$f_{\times}^{2} = f_{P}^{2}(F_{P3}^{2})f_{Q}^{0}(F_{Q1}^{0}) + f_{P}^{1}(F_{P2}^{1})f_{Q}^{1}(F_{Q2}^{1}) + f_{P}^{0}(F_{P1}^{0})f_{Q}^{2}(F_{Q3}^{2}) = 20\cdot 3 + 30\cdot 3 + 12\cdot 1 = 162$$
$$= 90F_{\times 4}^{2} + 72F_{\times 3}^{2},$$

$$f_{\times}^{3} = f_{P}^{3}(F_{P12}^{3})f_{Q}^{0}(F_{Q1}^{0}) + f_{P}^{2}(F_{P3}^{2})f_{Q}^{1}(F_{Q2}^{1}) + f_{P}^{1}(F_{P2}^{1})f_{Q}^{2}(F_{Q3}^{2}) = 1\cdot 3 + 20\cdot 3 + 30\cdot 1 = 93$$
$$= 3F_{\times 12}^{3} + 90F_{\times 6}^{3},$$

$$f_\times^4 = f_P^3(F_{P12}^3)f_Q^1(F_{Q2}^1) + f_P^2(F_{P3}^2)f_Q^2(F_{Q3}^2) = 1\cdot 3 + 20\cdot 1 = 23$$
$$= 3F_{\times 24}^4 + 20F_{\times 9}^4,$$

$$f_\times^5 = f_P^3(F_{P12}^3)f_Q^2(F_{Q3}^2) = 1\cdot 1 = 1 = P_{\times 36}^5.$$

Thus, the product of an icosahedron by a triangle is a 5 - polytope including 3 icosahedral prisms (11) and 20 3*3 - angular prismahedrons (17).

The structural formula of the polytope P_{36}^5 has the form

$$P_{36}^5\{20F_9^4[6F_6^3(3F_4^2,2F_3^2)],3F_{24}^4[20F_6^3(3F_4^2,2F_3^2),2F_{12}^3(20F_3^2)]\}. \tag{22}$$

We can call the polytope (22) an icosa-polytopic prismahedron.

The Product of a 4 - Simplex by a Triangle

According to theorem 1 and ratios (5) for the numbers of elements of different dimension in the product $P_5^3[5F_4^3(4F_3^2)]\times Q_3^2$ we have

$$f_\times^0 = f_P^0 f_Q^0 = 5\cdot 3 = 15$$

$$f_\times^1 = f_P^1 f_Q^0 + f_P^0 f_Q^1 = 10\cdot 3 + 5\cdot 3 = 45,$$

$$f_\times^2 = f_P^2(F_{P3}^2)f_Q^0(F_{Q1}^0) + f_P^1(F_{P2}^1)f_Q^1(F_{Q2}^1) + f_P^0(F_{P1}^0)f_Q^2(F_{Q3}^2) = 10\cdot 3 + 10\cdot 3 + 5\cdot 1 = 65$$
$$= 30F_{\times 4}^2 + 35F_{\times 3}^2,$$

$$f_\times^3 = f_P^3(F_{P4}^3)f_Q^0(F_{Q1}^0) + f_P^2(F_{P3}^2)f_Q^1(F_{Q2}^1) + f_P^1(F_{P2}^1)f_Q^2(F_{Q3}^2) = 5\cdot 3 + 10\cdot 3 + 10\cdot 1 = 55$$
$$= 15F_{\times 4}^3 + 40F_{\times 6}^3,$$

$$f_\times^4 = f_P^4(F_{P5}^4)f_Q^0(F_{Q1}^0) + f_P^3(F_{P4}^3)f_Q^1(F_{Q2}^1) + f_{P3}^2(F_P^2)f_Q^2(F_{Q3}^2) = 1\cdot 3 + 5\cdot 3 + 10\cdot 1 = 28$$
$$= 3F_{\times 5}^4 + 15F_{\times 8}^4 + 10F_{\times 9}^4,$$

$$f_{\times}^{5}=f_{P}^{4}(F_{P5}^{4})f_{Q}^{1}(F_{Q2}^{1})+f_{P}^{3}(F_{P4}^{3})f_{Q}^{2}(F_{Q3}^{2})=1\cdot 3+5\cdot 1=8$$
$$=3F_{\times 10}^{5}+5F_{\times 12}^{5},$$

$$f_{\times}^{6}=f_{P}^{4}(F_{P5}^{4})f_{Q}^{2}(F_{Q3}^{2})=1\cdot 1=1=P_{\times 15}^{6}.$$

Thus, the product of a 4 - simplex by a triangle is a 6-polytope including 3 5 – simplex - polytopic prisms (13) and 5 tetrahedral prismahedrons (18).

The structural formula of the polytope P_{15}^{6} has the form

$$P_{15}^{6}\{3F_{10}^{5}[5F_{8}^{4}[4F_{4}^{3}(3F_{4}^{2},2F_{3}^{2}),2F_{4}^{3}(4F_{3}^{2})],2F_{5}^{4}[5F_{4}^{3}(4F_{3}^{2})]],$$
$$5F_{12}^{5}[4F_{9}^{4}[6F_{6}^{3}(3F_{4}^{2},2F_{3}^{2}),3F_{8}^{4}[4F_{6}^{3}(3F_{4}^{2},2F_{3}^{2}),2F_{4}^{3}(4F_{3}^{2})]]\}. \quad (23)$$

We can call the polytope (23) a 6 – simplex - polytopic prismahedron. The image of this polytope is shown in Figure 12.

The Product of a 4 – Cross - Polytope by a Triangle

According to theorem 1 and ratios (5) for the numbers of elements of different dimension in the product $P_{8}^{3}[16F_{4}^{3}(4F_{3}^{2})]\times Q_{3}^{2}$ we have

$$f_{\times}^{0}=f_{P}^{0}f_{Q}^{0}=8\cdot 3=24,$$

Figure 12. The 6 – simplex - polytopic prismahedron

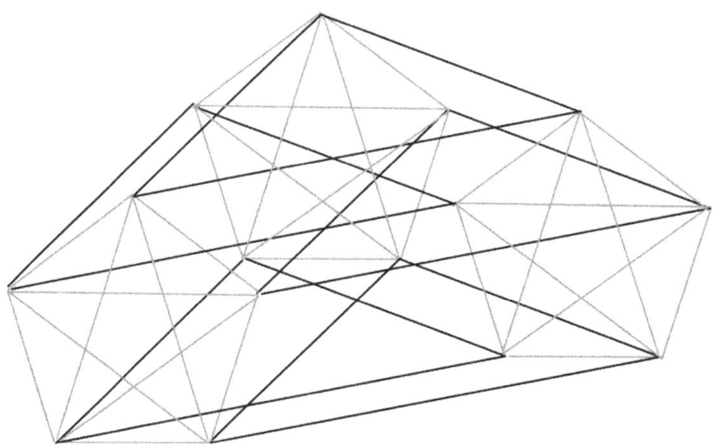

$$f_\times^1 = f_P^1 f_Q^0 + f_P^0 f_Q^1 = 24 \cdot 3 + 8 \cdot 3 = 96,$$

$$f_\times^2 = f_P^2(F_{P3}^2) f_Q^0(F_{Q1}^0) + f_P^1(F_{P2}^1) f_Q^1(F_{Q2}^1) + f_P^0(F_{P1}^0) f_Q^2(F_{Q3}^2) = 32 \cdot 3 + 24 \cdot 3 + 8 \cdot 1 = 176$$
$$= 72F_{\times 4}^2 + 104F_{\times 3}^2,$$

$$f_\times^3 = f_P^3(F_{P4}^3) f_Q^0(F_{Q1}^0) + f_P^2(F_{P3}^2) f_Q^1(F_{Q2}^1) + f_P^1(F_{P2}^1) f_Q^2(F_{Q3}^2) = 16 \cdot 3 + 32 \cdot 3 + 24 \cdot 1 = 168$$
$$= 48F_{\times 4}^3 + 120F_{\times 6}^3,$$

$$f_\times^4 = f_P^4(F_{P8}^4) f_Q^0(F_{Q1}^0) + f_P^3(F_{P4}^3) f_Q^1(F_{Q2}^1) + f_{P3}^2(F_P^2) f_Q^2(F_{Q3}^2) = 1 \cdot 3 + 16 \cdot 3 + 32 \cdot 1 = 83$$
$$= 51F_{\times 8}^4 + 32F_{\times 9}^4,$$

$$f_\times^5 = f_P^4(F_{P8}^4) f_Q^1(F_{Q2}^1) + f_P^3(F_{P4}^3) f_Q^2(F_{Q3}^2) = 1 \cdot 3 + 16 \cdot 1 = 19$$
$$= 3F_{\times 16}^5 + 16F_{\times 12}^5,$$

$$f_\times^6 = f_P^4(F_{P8}^4) f_Q^2(F_{Q3}^2) = 1 \cdot 1 = 1 = P_{\times 24}^6.$$

Thus, the product of a 4 – cross - polytope by a triangle is a 6-polytope including 3 5 – cross -polytopic prisms (10) and 16 tetrahedral prismahedrons (3).

The structural formula of the polytope P_{24}^6 has the form

$$P_{24}^6\{3F_{16}^5[16F_8^4[4F_6^3(3F_4^2, 2F_3^2), 2F_4^3(4F_3^2)], 2F_8^4[16F_4^3(4F_3^2)]],$$
$$16F_{12}^5[4F_9^4[6F_6^3(3F_4^2, 2F_3^2), 3F_8^4[4F_6^3(3F_4^2, 2F_3^2), 2F_4^3(4F_3^2)]]\}.$$

We can call the polytope a 6 - cross - polytopic prismahedron.

The General Structural Formula for the Product of Polytopes by a Triangle

If we pay attention to a polytope taking into account its dimensions and the number of vertices and the number of its facets and the number of vertices of every facet, we can prove the general rule of changing these parameters of a polytope for its multiplication by a triangle.

Theorem 3

Let's assume that a convex polytope of dimension *n,* with the number of vertices *m*, has a number of facets ν with vertex number *k* for every facet. Then its product by a triangle is defined by the equation

$$P_m^n(\nu F_k^{n-1}) \times Q_3^2 = P_{3m}^{n+2}(\nu F_{3k}^{n+1}, 3F_{2m}^{n+1}). \tag{24}$$

Proof

When multiplying the polytope P_m^n by the triangle Q_3^2 the number of vertices of the polytope is multiplied by 3, since the triangle has 3 edges and each edge at this multiplication translates polytope vertices in its direction. At that the dimension of the polytope naturally, according to the definition (Ziegler, 1995), is increased by two. Thus, we'll get a polytope P_{3m}^{n+2}. If we multiply the polytope by the triangle, its facets F_k^{n-1} are multiplied by the triangle also. The number of these facets of the product is equal to the number of facets of the polytope P_m^n, i.e. it is equal to ν. In addition, at the multiplication by the triangle the polytope P_m^n is multiplied by each edge (i.e. multiplication by Q_2^1 3 times). Therefore, 3 facets F_{2m}^{n+1} more appear as facets of the polytope P_{3m}^{n+2}. It is easy to make sure that all the considered above products of the polytopes by the triangle satisfy the equality (24).

Theorem 3 is proved.

Corollary: If the polygon Q_t^2 with the number of vertices *t* is the polytope factor in (24), the equation (24) can be rewritten as

$$P_m^n(\nu F_k^{n-1}) \times Q_t^2 = P_{tm}^{n+2}(\nu F_{tk}^{n+1}, \mathrm{t}\, F_{2m}^{n+1}). \tag{25}$$

As an example, the result of the product of a tetrahedron by a square is given in Figure 13.

Figure 13. The product of a tetrahedron by a square

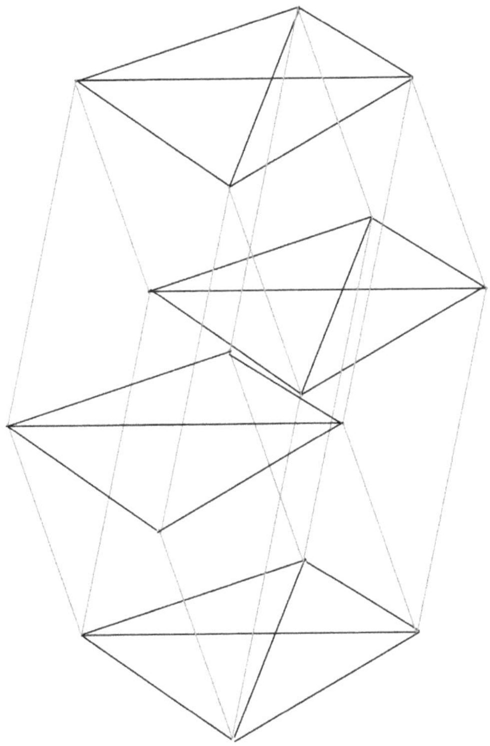

THE PRODUCTS OF POLYTOPES BY A TETRAHEDRON

The Product of a Tetrahedron by a Tetrahedron

According to theorem 1 and ratios (5) for the numbers of elements of different dimension in the product $P_4^3(4F_3^2) \times Q_3^2(4F_3^2)$ we have

$$f_\times^0 = f_P^0 f_Q^0 = 4 \cdot 4 = 16,$$

$$f_\times^1 = f_P^1 f_Q^0 + f_P^0 f_Q^1 = 6 \cdot 4 + 4 \cdot 6 = 48,$$

$$f_\times^2 = f_P^2(F_{P3}^2) f_Q^0(F_{Q1}^0) + f_P^1(F_{P2}^1) f_Q^1(F_{Q2}^1) + f_P^0(F_{P1}^0) f_Q^2(F_{Q3}^2) = 4 \cdot 4 + 6 \cdot 4 + 4 \cdot 4 = 68$$
$$= 36F_{\times 4}^2 + 32F_{\times 3}^2,$$

$$f_{\times}^{3} = f_{P}^{3}(F_{P4}^{3})f_{Q}^{0}(F_{Q1}^{0}) + f_{P}^{2}(F_{P3}^{2})f_{Q}^{1}(F_{Q2}^{1}) + f_{P}^{1}(F_{P2}^{1})f_{Q}^{2}(F_{Q3}^{2}) + f_{P}^{0}(F_{P1}^{0})f_{Q}^{3}(F_{Q4}^{3}) = 1\cdot 4 + 4\cdot 6 + 6\cdot 4 + 4\cdot 1 = 56 = 8F_{\times 4}^{3} + 48F_{\times 6}^{3},$$

$$f_{\times}^{4} = f_{P}^{3}(F_{P4}^{3})f_{Q}^{1}(F_{Q1}^{1}) + f_{P}^{2}(F_{P3}^{2})f_{Q}^{2}(F_{Q3}^{2}) + f_{P}^{1}(F_{P2}^{1})f_{Q}^{3}(F_{Q4}^{3}) = 1\cdot 6 + 4\cdot 4 + 6\cdot 1 = 28 = 12F_{\times 8}^{4} + 16F_{\times 9}^{4},$$

$$f_{\times}^{5} = f_{P}^{3}(F_{P4}^{3})f_{Q}^{2}(F_{Q3}^{2}) + f_{P}^{2}(F_{P3}^{2})f_{Q}^{3}(F_{Q4}^{3}) = 1\cdot 4 + 4\cdot 1 = 8 = 8F_{\times 12}^{5},$$

$$f_{\times}^{6} = f_{P}^{3}(F_{P4}^{3})f_{Q}^{2}(F_{Q4}^{3}) = 1\cdot 1 = 1 = P_{\times 16}^{6}.$$

Thus, the product of a tetrahedron by a tetrahedron is a 6-polytope including 8 tetrahedral prismahedrons (18).

The short structural formula of the polytope P_{16}^{6} has the form

$$P_{16}^{6}(8F_{12}^{5}). \tag{26}$$

The image of this polytope is shown in Figure 14.

We can call the polytope (26) a 6 - complex of tetrahedral prismahedrons.

The Product of an Octahedron by a Tetrahedron

According to theorem 1 and ratio (5) for the numbers of elements of different dimension in the product $P_{6}^{3}(8F_{P3}^{2})\times Q_{4}^{3}(4F_{Q3}^{2})$ we have

$$f_{\times}^{0} = f_{P}^{0}f_{Q}^{0} = 6\cdot 4 = 24,$$

$$f_{\times}^{1} = f_{P}^{1}f_{Q}^{0} + f_{P}^{0}f_{Q}^{1} = 12\cdot 4 + 6\cdot 6 = 84,$$

$$f_{\times}^{2} = f_{P}^{2}(F_{P3}^{2})f_{Q}^{0}(F_{Q1}^{0}) + f_{P}^{1}(F_{P2}^{1})f_{Q}^{1}(F_{Q2}^{1}) + f_{P}^{0}(F_{P1}^{0})f_{Q}^{2}(F_{Q3}^{2}) = 8\cdot 4 + 12\cdot 6 + 6\cdot 4 = 128 = 72F_{\times 4}^{2} + 56F_{\times 3}^{2},$$

Figure 14. The product of a tetrahedron by a tetrahedron

$f_{\times}^{3} = f_{P}^{3}(F_{P6}^{3})f_{Q}^{0}(F_{Q1}^{0}) + f_{P}^{2}(F_{P3}^{2})f_{Q}^{1}(F_{Q2}^{1}) + f_{P}^{1}(F_{P2}^{1})f_{Q}^{2}(F_{Q3}^{2}) + f_{P}^{0}(F_{P1}^{0})f_{Q}^{3}(F_{Q4}^{3}) = 1\cdot 4 + 8\cdot 6 + 12\cdot 4 + 6\cdot 1 = 106 = 6F_{\times 4}^{3} + 100F_{\times 6}^{3}$,

$f_{\times}^{4} = f_{P}^{3}(F_{P6}^{3})f_{Q}^{1}(F_{Q1}^{1}) + f_{P}^{2}(F_{P3}^{2})f_{Q}^{2}(F_{Q3}^{2}) + f_{P}^{1}(F_{P2}^{1})f_{Q}^{3}(F_{Q4}^{3}) = 1\cdot 6 + 8\cdot 4 + 12\cdot 1 = 50 = 12F_{\times 8}^{4} + 32F_{\times 9}^{4} + 6F_{\times 12}^{4}$,

$f_{\times}^{5} = f_{P}^{3}(F_{P6}^{3})f_{Q}^{2}(F_{Q3}^{2}) + f_{P}^{2}(F_{P3}^{2})f_{Q}^{3}(F_{Q4}^{3}) = 1\cdot 4 + 8\cdot 1 = 12 = 8F_{\times 12}^{5} + 4F_{\times 18}^{5}$,

$f_{\times}^{6} = f_{P}^{3}(F_{P6}^{3})f_{Q}^{2}(F_{Q4}^{3}) = 1\cdot 1 = 1 = P_{\times 24}^{6}$.

Thus, the product of an octahedron by a tetrahedron is a 6 - polytope including 8 tetrahedral prismahedrons (18) and 4 octahedral polytopic prisms (19).

The short structural formula of the polytope P^6_{24} has the form

$$P^6_{24}(4F^{\ 5}_{18}, 8F^5_{12}). \tag{27}$$

The image of this polytope is shown in Figure 15.

We can call the polytope P^6_{24} a 6 - complex of tetrahedral prismahedrons and octahedral polytopic prisms.

Figure 15. The product of an octahedron by a tetrahedron

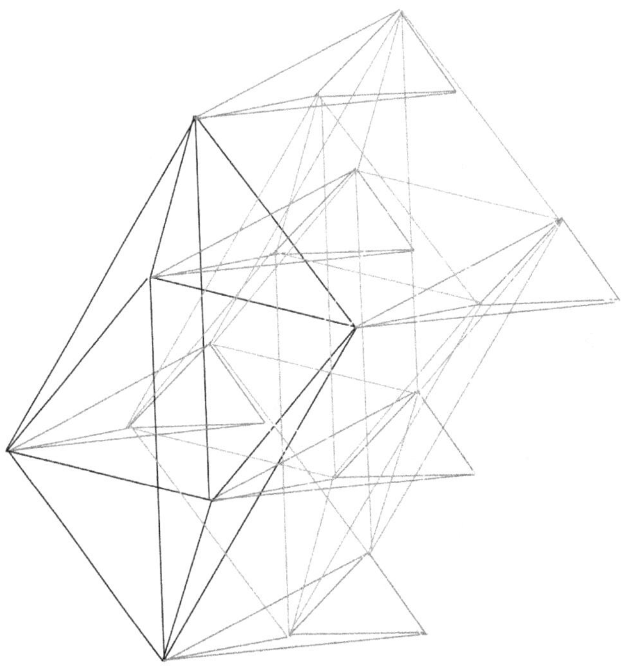

THE PRODUCNT OF POLYTOPES WITH UNIFORM FACETS

Comparing (26) and (27) with equalities defined by theorems 1, 2 we note their general structure. This allows us to formulate and prove a general theorem on the product of polytopes with uniform facets.

Theorem 4

Let's assume that we have a convex polytope *P* of dimension *n,* with the number of vertices *m*, which has ν facets with the number of vertices *k* for each facet. Let's also set a convex polytope *Q* of dimension π, with the number of vertices *l*, having μ facets with the number of vertices λ for every facet. Then the product of these polytopes is defined by equation

$$P^{n}_{m}(\nu F^{n-1}_{k}) \times Q^{\pi}_{l}(\mu F^{\pi-1}_{\lambda}) = P^{n+\pi}_{lm}(\nu F^{n+\pi-1}_{lk}, \mu F^{n+\pi-1}_{\lambda m}). \tag{28}$$

Proof

For multiplication of the polytope P^{n}_{m} by the polytope Q^{π}_{l} the number of the vertices of polytope *P* is multiplied by the number of the vertices of polytope *Q* since each vertex of the polytope *P* at multiplication will correspond to all vertices of the polytope *Q*. The dimension of the product polytope according to the definition (Ziegler, 1995) is the sum of the dimensions of polytopes-factors, i.e. we receive the polytope $P^{n+\pi}_{lm}$ as the result of the product. Multiplication of the polytope *P* by the polytope *Q* is accompanied also by multiplication of the polytope *P* facets by the polytope *Q*. That is why the result of this multiplication will be the facets of the polytopes product defined according to the same rules, i.e. $F^{n+\pi-1}_{lk}$. The number of these facets is equal to the number of facets of the polytope *P*, i.e. ν. Besides, in the product of the polytopes *P* and *Q* there are facets formed by the product of the facets of polytope *Q* by polytope *P*. These are facets $F^{n+\pi-1}_{lk}$, their number is equal to the number of facets of the polytope *Q*, i.e. μ. The theorem 4 is proved.

Let's use theorem 4 for the product of an octahedron by a tetrahedron

$$P_6^3(8F_{P3}^2)\times Q_4^3(4F_{Q3}^2)=P_{24}^6(4F_{18}^{\ 5},8F_{12}^5).$$

The received formula coincides with (27) got after detailed derivation.

It's easy to verify that all the previous formulas about polytopes product are particular cases of the formula (28). Theorem 4 allows to write down at once the result of the product of any convex polytopes with uniform faces without detailed derivation of this result. Let's build the multiplication table of existing polytopes of dimensions 3 and 4. For comfort of presentation the table is divided into a few tables (tables 1 – 4) to cover the products of all above-mentioned polytopes. The structure of new formed facets can be defined according to already mentioned formulas. The structural formulas in the table are given in a short form.

Table 1. Multiplication table of polytopes

Polytopes	Tetrahedron	Octahedron
Tetrahedron	$P_{16}^6(8F_{12}^5)$	
Octahedron	$P_{24}^6(4F_{18}^5,8F_{12}^5)$	$P_{36}^6(16F_{18}^5)$
Cube	$P_{32}^6(6F_{16}^5,4F_{12}^5)$	$P_{48}^6(6F_{24}^{5^0},8F_{24}^{5^{00}})$
Dodecahedron	$P_{80}^6(12F_{20}^5,8F_{60}^5)$	$P_{120}^6(12F_{130}^5,8F_{60}^5)$
Icosahedron	$P_{48}^6(20F_{12}^5,4F_{36}^5)$	$P_{72}^6(20F_{18}^5,8F_{36}^5)$
4-simplex	$P_{20}^7(5F_{16}^5,4F_{15}^5)$	$P_{30}^7(5F_{24}^{6^0},8F_{15}^6)$
4-cube	$P_{64}^7(8F_{32}^6,4F_{48}^6)$	$P_{96}^7(8F_{48}^{6^0},8F_{48}^{6^{00}})$
4-cross-polytope	$P_{32}^7(16F_{16}^6,4F_{24}^6)$	$P_{48}^7(16F_{24}^{6^0},8F_{24}^{6^{00}})$

$F_{24}^{5^0}=F_4^2\times F_6^3$, $F_{24}^{5^{00}}=F_8^3\times F_3^2$, $F_{24}^{6^0}=F_4^3\times F_6^3$, $F_{24}^{6^{00}}=F_8^4\times F_3^2$, $F_{48}^{6^0}=F_8^3\times F_6^3$, $F_{48}^{6^{00}}=F_{16}^4\times F_3^2$.

Table 2. Multiplication table of polytopes

Polytopes	Cube	Dodecahedron
Cube	$P_{64}^{6}(12F_{32}^{5})$	
Dodecahedron	$P_{160}^{6}(12F_{40}^{5},6F_{80}^{5})$	$P_{400}^{6}(24F_{100}^{5})$
Icosahedron	$P_{96}^{6}(20F_{24}^{5},6F_{48}^{5})$	$P_{240}^{6}(20F_{60}^{5^0},12F_{60}^{5^{00}})$
4-simplex	$P_{40}^{7}(5F_{32}^{6^0},6F_{20}^{6})$	$P_{160}^{7}(16F_{80}^{6},12F_{40}^{6})$
4-cube	$P_{128}^{7}(6F_{32}^{6^0},8F_{64}^{6})$	$P_{320}^{7}(8F_{160}^{6},12F_{80}^{6})$

$$F_{32}^{6^0} = F_4^3 \times F_8^3,\ F_{32}^{6^{00}} = F_8^4 \times F_4^2,\ F_{60}^{5^0} = F_3^2 \times F_{20}^3,\ F_{60}^{5^{00}} = F_{12}^3 \times F_5^2.$$

Table 3. Multiplication table of polytopes

Polytopes	Icosahedron	4-simplex
Icosahedron	$P_{144}^{6}(40F_{36}^{5})$	
4-simplex	$P_{60}^{7}(5F_{48}^{6},20F_{15}^{6})$	$P_{25}^{8}(10F_{20}^{7})$
4-cube	$P_{192}^{7}(8F_{96}^{6},20F_{48}^{6})$	$P_{80}^{8}(5F_{64}^{7},8F_{40}^{7})$
4-cross-polytope	$P_{96}^{7}(16F_{48}^{6},20F_{24}^{6})$	$P_{40}^{8}(5F_{32}^{7},16F_{20}^{7})$

Table 4. Multiplication table of polytopes

Polytopes	4 –cube	4 –cross-polytope
4 –cube	$P_{256}^{8}(16F_{128}^{7})$	
4 – cross-polytope	$P_{128}^{8}(16F_{64}^{7^0},8F_{64}^{7^{00}})$	$P_{64}^{8}(32F_{32}^{7})$

$$F_{64}^{7^0} = F_4^3 \times F_{16}^4,\ F_{64}^{7^{00}} = F_8^4 \times F_8^3.$$

THE INCIDENCE VALUES IN THE POLYTOPIC PRISMAHEDRONS

The incidence of elements of the boundary complexes of polytopic prisms is also preserved for polytopic prismahedrons. For many of them Theorem 2 and formula (5) of the Chapter 2 are satisfied. For example, for a polytopic prismahedron that is a product of a square and a tetrahedron (Figure 13), the incidence values of elements of a boundary complex of different dimensions have the form

$$k_{b(i=0)}(5) = 5, k_{b(i=1)}(5) = 4, k_{b(i=2)}(5) = 3, k_{b(i=3)}(5) = 2, k_{b(i=4)}(5) = 1.$$

This completely coincides with Theorem 2 and formula (5) of Chapter 2.

For the polytopic prismahedron, which is the product of a tetrahedron on a tetrahedron (Figure 14), the incidence values of elements of a boundary complex of different dimensions have the form

$$k_{b(i=0)}(6) = 6, k_{b(i=1)}(6) = 5, k_{b(i=2)}(6) = 4, k_{b(i=3)}(6) = 3, k_{b(i=4)}(6) = 2, k_{b(i=5)}\left(6\right) = 1.$$

This also agrees with Theorem 2 and formula (5) of Chapter 2.

However, for polytopic prismahedron in which a pyramid participates (or an octahedron that can be considered as a double pyramid), there is a difference from Theorem 2 and the formula (5) of Chapter 2. This difference was already observed when considering the product of the Egyptian pyramid on a one - dimensional segment (Figure 3). The incidence of vertices in this case is greater than the value determined by Theorem 2 and formula (5) of Chapter 2. For example, for a polytopic prismahedron that is the product of an octahedron per triangle (Figure 10), the incidence values of the elements of the boundary complex have the form

$$k_{b(i=0)}(5) = 6, k_{b(i=1)}(5) = 4, k_{b(i=2)}(5) = 3, k_{b(i=3)}(5) = 2, k_{b(i=4)}(5) = 1.$$

The incidence of vertices in this case is greater than the dimension of the polytopic prismahedron.

The participation of cross-polytopes as elements of the boundary complex of polytopic prismahedrons significantly changes the incidence of the elements of the boundary complex, as this has already been noted when considering polytopic prisms. The influence of cross - polytopes on the incidence of elements of the boundary complex in complicated polytopic prismahedrons has yet to be investigated.

HIERARCHICAL AND TRANSLATIONAL FILLING SPACES

In monograph of Zhizhin (Zhizhin, 2017) it showed, that many molecules have a dimension greater than three if they are described in the form of geometric convex figures. Many of them have tetrahedral and octahedral coordination. But any substance is not a separate molecule, but a collection of molecules. This raises the problem of filling spaces with molecules, i.e. convex geometric figures of different dimensions. The addition of other molecules (indifferent or not) leads to the formation of a cluster. There are various patterns in the formation of clusters (Lord, Mackay & Ranganathan, 2006; Ilyushin, 2003). You can distinguish different levels of organization in clusters. From a geometric point of view, this amounts to a hierarchical filling of the space (Zhizhin, 2012). However, the hierarchical filling of space is always limited by the size of some area of space, primarily because when distance from the original molecule the distance between the particles (or points) increases significantly (Zhizhin, 2014). This inevitably leads to a weakening of the bonds between molecules and, as a consequence, to the detachment of molecules. Therefore, in order to further fill of the space, it is necessary to proceed to space translation a certain set of atoms or molecules (cluster). If a molecule in geometric relation is a multidimensional formation, then the problem arises of filling the space by means of translation by a multidimensional convex body. If a molecule or molecular formation is a multidimensional body, then this body must be translated from the coordinates of a multidimensional space. Moreover, in order for such a translation to be possible, the body itself must have a certain structure, i.e. to have selected directions along the coordinates in itself. Then translation of this body in the selected directions will not cause a violation of the body shape. The products of polytopes considered in this chapter just create forms in which there are selected directions along the coordinates of a multidimensional space. They are created by the very mechanism of the product polytopes. The mathematical apparatus developed in the chapter for determining the shape of a body that is a product of polytopes can be applied to determining the shape of bodies not listed in Tables 1 - 4. The diffraction pattern of the quasicrystal, shown in the Figure 1 in Chapter 4, reveals the nanostructure of the intermetallic compound $Al_{72}Ni_{20}Co_8$. You can see similar structures on other substances (Shechtman et al., 1984; Mukhopadhyay et al., 1993; Zhang & Kelton, 1993). The diffraction pattern clearly shows the system of parallel lines formed as result of the translation of molecules along different directions of coordinates (Shevchenko, Zhizhin & Mackay, 2013). At the same time, the lines formed as result of the hierarchical filling of space are visible. In fact, from each point one can see the formation of lines as a result of the hierarchical filling of space (if we neglect the influence of the experimental features on the diffraction pattern - the axial symmetry caused by the axial symmetry of the electron beam directed to the sample of matter). Just as there is an expansion of space at each point of space in the model of the expansion of the Universe (Silk, 1980).

REFERENCES

Alexandrov, P. S. (1975). *Introduction to homological dimension theory and general combinatorial topology*. Moscow: Science.

Delone, B. (1961). Proof of the main theorem of the theory of stereohedrons. *Reports of the Academy of Sciences of the USSR, 138*(6), 1270–1272.

Delone, B., & Sandakova, N. (1961). Theory of stereohedrons. *Proceedings of the Mathematical Institute. V.A. Steklov, 64*, 28-51.

Fomenko, A. T. (1992). *Visual geometry and topology. Mathematical images in the real world*. Moscow: Moscow State University.

Ilyushin, G. D. (2003). *Modeling of self - organization processes in crystal-forming systems*. Moscow: URSS.

Lord, E. A., Mackay, A. L., & Ranganathan, S. (2006). *New geometry for new materials*. Cambridge: Cambridge University Press.

Mukhopadhyay, N.K., Weatherly, G.C., Lloyd, D.J. & Embury, J.D. (1993). Diffraction studies of icosahedral phases in $Al_{70}Fe_{20}W_{10}$. *Journal of Non-Crystalline Solids, 153-154*, 1193-1197.

Panina, G. Y. (2006). The Algebra of Polyhedrons.

Pontryagin, L. S. (1976). *The foundations of combinatorial topology*. Moscow: Science.

Poincare A. (1895). Analysis situs. *J. de é Ecole Polytechnique, 1,* 1 – 121.

Robertson, S. A. (1984). *Polytopes and Symmetry.* Cambridge: Cambridge University Press.

Shechtman, D., Blech, I., Gratias, D., & Cahn, J. W. (1984). Metallic phase with longerange orientational order and no translational symmetry. *Physical Review Letters, 53*(20), 1951–1953. doi:10.1103/PhysRevLett.53.1951

Silk, J. (1980). *The big bang. The creation and evolution of the Universe.* San Francisco: W.H. Freeman and Company.

Shevchenko, V. Ya., Zhizhin, G. V., & Mackay, A. (2013). On the structure of the quasi-crystals in the high dimension space. In M. V. Diudea (Ed.), *Diamonds and related nanostructures* (pp. 311–320). Dordrecht: Springer. doi:10.1007/978-94-007-6371-5_17

Snegirev, I. L., & Frantsev, Yu. P. (1938). *Ancient Egypt.* Leningrad: OGIZ.

Zang, X., & Kelton, K. (1993). High-order crystal approximant alloys $Ti_{54}Zr_{26}Ni_{20}$. *Journal of Non-Crystalline Solids, 153-154*, 114–118. doi:10.1016/0022-3093(93)90325-R

Zhizhin, G. V. (2012, October). Hierarchical filling of spaces with polytopes. *Paper presented at "7th St.- Petersburg meeting of Nobel Prize laureates,* St. Petersburg, Russia.

Zhizhin, G. V. (2014). *World – 4D.* St. – Petersburg: Polytechnic Service.

Zhizhin, G. V. (2015, November). Polytopic prismahedrons – fundamental regions of the n-dimension nanostructures. *Paper presented at The International conference "Nanoscience in Chemistry, Physics, Biology and Mathematics,"* Cluj – Napoca, Romania.

Zhizhin G.V. (2017a). N-prismahedrons and their dual polytopes. *International Journal Chemical Modeling, 5*(4).

Zhizhin, G. V. (2017b). *Chemical Compounds Structures and the Higher Dimension of Molecules: Emerging Research and Opportunities.* Hershey, PA: IGI Global.

Ziegler, G. (1995). *Lectures on polytopes.* New York: Springer. doi:10.1007/978-1-4613-8431-1

KEY TERMS AND DIFINITIONS

5 – Cross – Prism: The product of 4 – cross – polytope by one dimension segment.

5 – Simplex – Prism: The product of 4 – simplex by one dimension segment.

6 – Complex – Polytopic Prismahedrons: The product of tetrahedron by the tetrahedron.

6 – Complex – Tetrahedral Prismahedrons and Octahedral Prismahedrons: The product octahedral by the tetrahedral.

6 – Simplex – Polytopic Prismahedron: The product of 4 – simplex by the triangle.

Cube – Polytopic Prismahedron: The product of cube by the triangle.

Dodeca - Polytopic Prismahedron: The product of dodecahedron by the triangle.

Dodecahedral Prism: The product of dodecahedral by one dimension segment.

Hierarchical and Translation Filling Spaces: At the same time hierarchical and translation filling of a multidimensional spaces.

Icosa – Polytopic Prismahedron: The product of icosahedron by the triangle.

Icosahedral Prism: The product of icosahedral by one dimension segment.

N*3 – Angular Prismahedron: The product of n – angle by the triangle.

Octahedral Prism: The product of octahedral by one dimension segment.

Octahedral Prismahedron: The product of octahedral by the triangle.

Polytopic Prismahedron: The product of polytope by one dimension segment.

Tetrahedral Prism: The product of tetrahedral by one dimension segment.

Tetrahedral Prismahedron: The product of tetrahedral by the triangle.

Triangular Prismahedron: The product of triangle by the triangle.

Chapter 5
Poly – Incident and Dual Polytopes

ABSTRACT

The incidence of elements of low dimension in convex regular polytopes of dimension n with respect to elements of higher dimension up to elements of dimension n - 1 is investigated. It is shown that polytopes are dual to polytopic prismahedrons form a new class of polytopes with simultaneously different values of the incidence of elements of low dimension to elements of higher dimension entering in the polytope. This new type of polytopes is called poly – incident polytopes. The existence of a previously unknown polytope consisting of one hundred tetrahedrons is established. All its constituent tetrahedrons are listed. The concept of a polytope with a factor structure whose vertices consist of polytopes of large dimension is introduced.

INTRODUCTION

In Chapter 4 it was described that the coefficient of incidence of the vertices of the Egyptian pyramid and polytopic prismahedrons with the participation of pyramids in their boundary complexes can have simultaneously several different values in the same polytope. It turns out that polytopic prismahedrons lead to polytopes with different incidence values not only of vertices, but also of other elements of boundary complexes and in the absence of pyramids.

DOI: 10.4018/978-1-5225-6968-8.ch005

Let's denote $b(i)$ - an element of dimension *i*; $k_{b(i),b(j)}$ - the value of the incidence of the element with dimension *i* in relation to the elements of dimension *j* (*j* > *i*). For regular polytopes because of their uniform values of the incidence $k_{b(i),b(j)}$ are constant for the whole polytope in all dimension range from 0 to *n* (*n* - dimension of the polytope). Obviously, that $k_{b(i),b(n)}$ = 1 for any $i < n$.

In a polygon there are:

$$b(0) \in 2b(1); b(1) \in b(2).$$

In a polyhedron there are:

$$b(3): b(0) \in k_{b(0),b(1)} b(1); k_{b(0),b(1)} = 3,4,5; b(1) \in 2b(2); b(2) \in b(3).$$

In four - dimensional polytopes relations of the incidence have the following values.

In a 4 - simplex:

$$b(0) \in 4b(1); b(1) \in 3b(2); b(2) \in 2b(3); b(3) \in b(4); b(0) \in 4b(3); b(0) \in 4b(2).$$

In a 4 - cube:

$$b(0) \in 4b(1); b(1) \in 3b(2); b(2) \in 2b(3); b(3) \in b(4); b(0) \in 3b(3); b(0) \in 6b(2); b(1) \in 3b(3).$$

In a 4 – cross - polytope:

$$b(0) \in 6b(1); b(1) \in 4b(2); b(2) \in 2b(3); b(3) \in b(4); b(0) \in 8b(3); b(0) \in 10b(2); b(1) \in 4b(3).$$

In the construction of dual polytopes to polytopes of dimension *n*, the incidence coefficients with respect to facets of polytopes $k_{b(i),b(n-1)}$. For correct four - dimensional polytopes, these incidence coefficients have, respectively, the values:

In a 4 – simplex:

$$k_{b(0),b(3)} = 4; k_{b(1),b(3)} = 3; k_{b(2)b(3)} = 2;$$

In a 4 – cube:

$$k_{b(0),b(3)} = 4; k_{b(1),b(3)} = 3; k_{b(2)b(3)} = 2;$$

In a 4 – cross – polytope:

$$k_{b(0),b(3)} = 8; k_{b(1),b(3)} = 4; k_{b(2)b(3)} = 2.$$

In semi - regular polytopes relations of incidence keep their form the same as in regular polytopes (Zhizhin, 2014a). But there different figures in one polytope may serve as elements $b(2)$, though all vertices of these semi - regular polytopes are superposed by motion. If in a polytope there are vertices which are not superposed by motion, then relations of incidence are variable in a polytope, for example, in a triangle prism. A prism can be considered a semi - regular polytope because it has two triangle faces and three of square faces.

For a deeper study of the geometry of polytopic prismahedron, one apply the analysis of the duality of polytopic prismehedron. The analysis of duality, as an analysis of the unity of opposites, is used in various fields of science: in mathematics, logic, philosophy, physics and chemistry (Bogdanov, 1989; Whitehead, 1990; Feyman, 1968; Hegel; 1998; Bucur and Delanu, 1972).

One will consider the polytope as dual to a given polytope of dimension n if the facets of a given polytope are replaced on the vertices of the polytope. Moreover, the vertices of a new polytope are connected by an edge if the corresponding facets have a common face of dimension n - 2 in the given polytope. Consider a rectangular prism with triangular bases. Can to introduce the centers of the planar faces of this prism and join the edges of those centers whose faces have a common edge. One will get a double pyramid (Figure 1). It is an irregular polyhedron dual to this prism.

In a double pyramid a two vertices are incident to three edges, and another two vertices are incident to four edges. The same picture will be if one take a pentagonal prism.

Two vertices of double pentagonal pyramid are incident to 5 edges, and 5 of the remaining vertices are incident to 4 edges (Figure 2).

Figure 1. The triangle prism and double pyramid dual to it

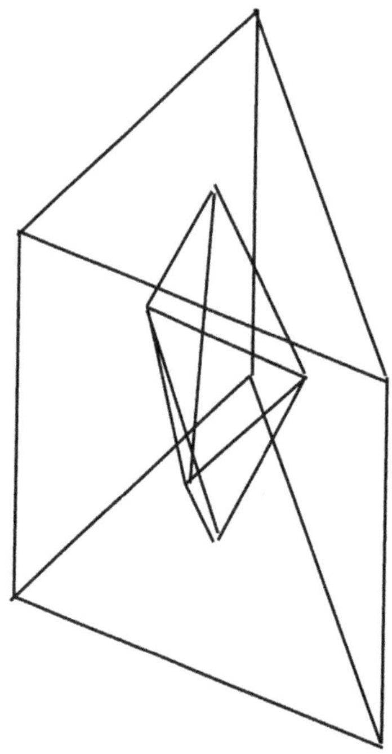

Figure 2. The pentagonal prism and a double pyramid dual to it

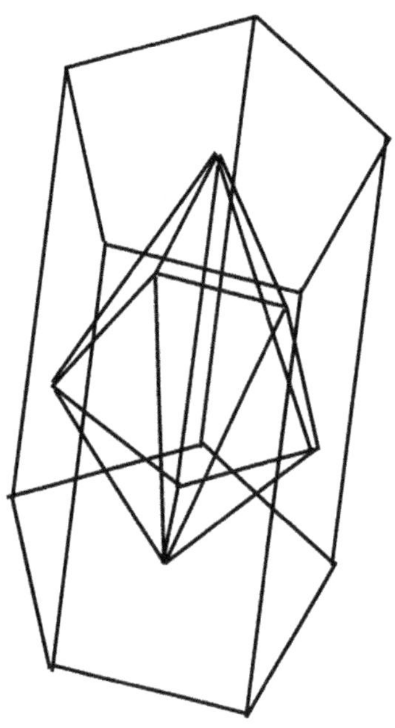

Further it will be shown that the polytopes dual to polytopes products have not only vertices with different values of incidence to the edges, but the edges with different values of incidence to three - dimensional figures. This new type of polytopes is called poly - incident polytopes.

POLYTOPE DUAL TO THE PRODUCT OF TWO TRIANGLES

In Chapter 4 the structural formula of the product of two triangles (1) was obtained, according to which this 4 - polytope has 9 vertices and 6 triangular prisms act as facets. This polytope has 18 edges, 9 squares and 6 triangles as two - dimensional elements. In a dual polytope owing to inverse of inclusion relation (Grunbaum, 1969) 6 vertices, 15 edges, 18 two - dimensional elements, 9 three - dimensional facets shall be. Each three - dimensional face has the number of two - dimensional faces equal to double ratio of the number of two - dimensional faces to the number of three - dimensional faces. Doubling happens due to the fact, that each flat face in a 4 -polytope simultaneously belongs to two three - dimensional faces, i.e. the number of flat faces belonging to three - dimensional faces of a dual polytope is $2 \cdot 18/9 = 4$. Therefore, the three -dimensional faces in a dual polytope are tetrahedrons, since only a tetrahedron has four flat faces. For the construction of a dual polytope let's use a sequence of actions that will be used further for construction of more complex polytopes. Let's denote vertices of the polytope in Figure 5 of Chapter 4 by symbols $a_1, ..., a_9$ (Figure 3).

Let's enumerate the triangular prismahedrons (facets) forming this polytope:

$$1)a_1a_2a_3a_7a_8a_9; 2)a_2a_3a_4a_5a_8a_9; 3)a_1a_2a_5a_6a_7a_8;$$
$$4)a_1a_2a_3a_5a_6a_4; 5)a_1a_6a_3a_4a_7a_9; 6)a_4a_6a_5a_7a_8a_9.$$

Since the edges of a dual polytope shall connect the centers of polytope facets adjoining each other, so, by introduced designations of facets, can to define neighboring facets having common flat faces by introduced designations of facets. Therefore, just these facets shall be connected by edge in a dual polytope (Zhizhin, 2016, 2017). Executing this analysis, one will come to the projection of a dual polytope on a two - dimensional plane (Figure 4).

In Figure 4 using the condition that each triangle shall be overlapped twice, can really find 9 tetrahedrons:

Figure 3.The triangular prismahedron

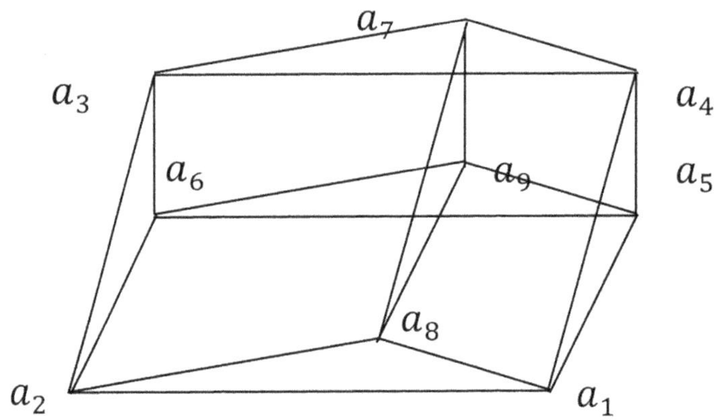

Figure 4. The polytope dual to the triangular prismahedron

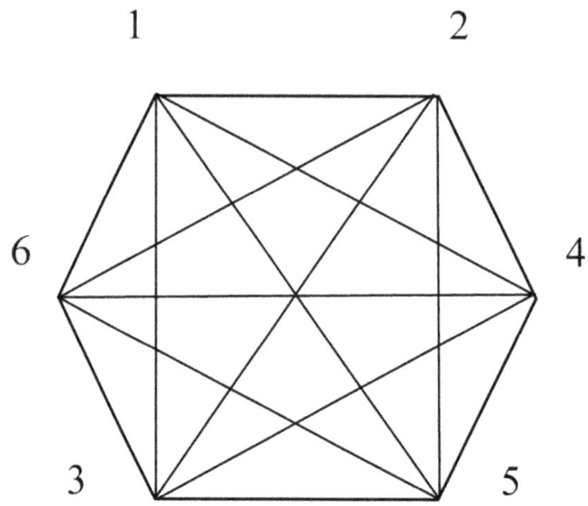

$T_1 = 1246, T_2 = 2345, T_3 = 1245, T_4 = 6435, T_5 = 1635, T_6 = 1362, T_7 = 1253, T_8 = 1645, T_9 = 6243.$
Let's enumerate the edges and define their belonging to the listed the tetrahedrons.

The edges: 61, 12, 24, 45, 35, 63, 13, 14, 34, 62, 25, 65, 15, 32, 64;

$$61 \in T_1, T_6, T_5, T_8; 12 \in T_1, T_3, T_6, T_7; 24 \in T_1, T_2, T_3, T_9; 45 \in T_2, T_4, T_8, T_3; 35 \in T_2, T_4, T_5, T_7;$$
$$63 \in T_6, T_4, T_5, T_9; 13 \in T_7, T_5, T_6; 14 \in T_1, T_3, T_8; 34 \in T_2, T_4, T_9; 62 \in T_1, T_6, T_9; 25 \in T_2, T_3, T_7;$$
$$65 \in T_4, T_5, T_8; 15 \in T_3, T_5, T_7, T_9; 32 \in T_2, T_6, T_7, T_9; 64 \in T_1, T_4, T_8, T_9.$$

Thus, there is a wonderful fact: the edges have different values of incidence to edges. The edges 13, 34, 14, 62, 25, 65 have value of incidence 3; the edges 61, 12, 24, 45, 35, 63, 15, 64, 32 have value of incidence 4. It occurs when the triangular prismahedron has all edges with equal values of incidence.

GENERAL ANALYSIS OF THE PRODUCT OF ANY TWO CONVEX POLYGONS

As a consequence from theorem 3 of Chapter 4 was received the structural formula (25) of the product of polytope P^n by a polygon. If both factors in (25) of Chapter 4 are polygons, the formula (25) take the form of

$$P_m^2 \times Q_t^2 = P_{tm}^4(mF_{t2}^3, tF_{2m}^3). \quad (1)$$

According to Chapter 4, can to call polytopes (1) the *mt* - angular prismahedrons.

Theorem 1

The *mt* - angular prismahedron is a 4 - polytope consisting of *m* the *t* - angular prisms and *t* the *m* - angular prisms.

Proof

As a result of the product of the *m* - angular polygon by the *t* – angular polygon, according to the definition of polytopes product (Ziegler, 1995), is a polytope with facets forms. These facets are the products of each side of the *m* - angular polygon by the *t* - angular polygon and, accordingly, the products of each sides of the *t* - angular polygon by the *m* - angular polygon. The first from the products give *m* the *t*- angular prisms, the second ones –

give *t* the *m* - angular prisms. Let's check by Euler - Pioncare's equation (2) in Chapter 2 (Pioncare, 1880) that the polytopes formed as a result of the product of polygons satisfy this equation for polytope dimension $n = 4$. In this case the number of elements of dimension 0 (vertices) owing to constructing polytopes is equal to *mt.* The number of one - dimension elements (edges) is composed of the number of edges issuing from all vertices with account of each edge passing through two vertices. Since 4 edges issue from each vertex owing to polygons product, the total number of the edges is $\frac{mt4}{2} = 2mt.$ The number of two - dimensional faces is composed of numbers *m* and *t,* correspondingly the *t* - angular and the *m* – angular bases of the prisms and the 4 - angular faces of the prisms *mt.*

Substituting the received values of numbers of elements of different dimension in Euler - Pioncare's equation

mt – 2mt +(m+t+mt) – (m +t)=0.

Thus, Euler – Pioncare`s equation is satisfied, which proves theorem 1.

Consequence. Owing to duality in a polytope dual to the *mt* - angular prismahedron, the number of vertices is equal to *m+t,* the number of edges is equal to *m+t+mt,* the number of the two - dimensional faces is equal to 2*mt,* the number of three - dimensional faces is equal to *mt.*

Each three - dimensional face has the number of two - dimensional faces equal to double ratio $\frac{2mt}{mt}$, i.e. 4. Thus, in the general case of the product of any two convex polytopes the tree -dimensional figure is a tetrahedron, since only a tetrahedron, as a three - dimensional figure, has 4 two - dimensional faces.

THE 4*3 - ANGULAR PRISMAHEDRON AND ITS DUAL POLYTOPE

In Chapter 4 the image of the 4*3 - angular prismahedron (Figure 4) was received. Let's denote vertices of this polytope $a_1,...,a_{12}$ (Figure 5).

The 4*3 - angular prismahedron includes three - dimensional figures: the triangular prisms

*Figure 5. The 4*3 - angular prismahedron*

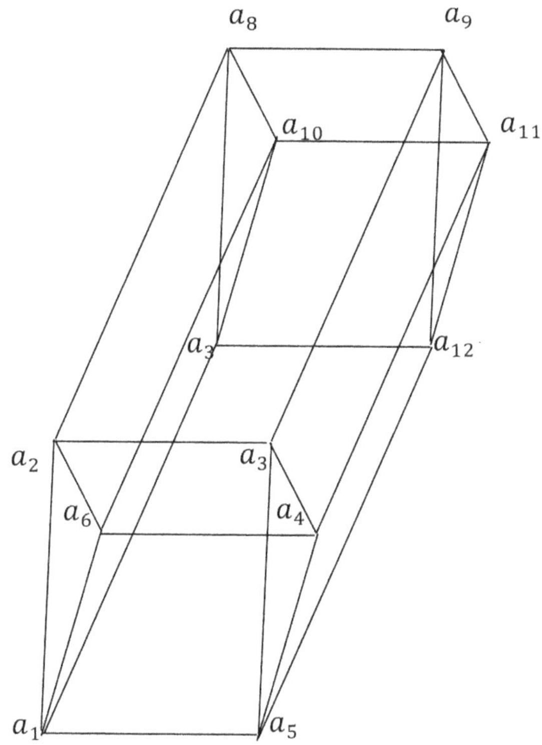

$1)a_1a_2a_6a_7a_8a_{10}, 2)a_3a_4a_5a_9a_{11}a_{12}, 3)a_1a_2a_6a_3a_4a_5, 4)a_7a_8a_{10}a_{11}a_{12};$

and the tetragonal prisms

$5)a_2a_6a_3a_4a_8a_{10}a_9a_{11}, 6)a_1a_6a_4a_5a_7a_{10}a_{11}a_{12}, 7)a_1a_2a_3a_5a_7a_8a_9a_{12}.$

For constructing dual polytope can to define contacts of the prisms by equal sets of symbols a_i. In the result can to receive a projection of dual polytope (Figure 6).

In polytope in Figure 6 one can single out 12 tetrahedrons with flat faces overlapped two times: 2564, 4527, 2537, 2563, 4561, 4571, 5731, 2746, 1536, 3627, 6714, 7136. This polytope have 19 edges: 17, 16, 15, 14, 13, 37, 36, 35, 32, 27, 26, 24, 25, 57, 56, 54, 47, 46, 67. From them edges 17, 14, 32, 24 have incidence value 3 with respect to tetrahedrons and the remaining

*Figure 6. Polytope dual to the 4*3 - angular prismahedron*

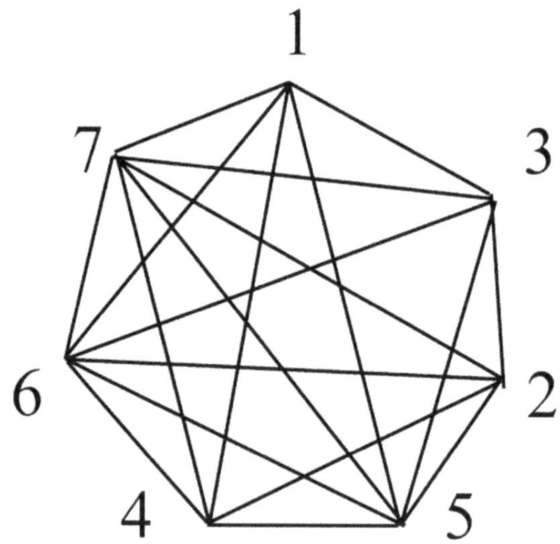

15 edges have incidence value 4. The vertices of dual polytope 3, 2, 1, 4 are incident to 5 edges and vertices 5, 6, 7 are incident to 6 edges.

THE 3*6 - ANGULAR PRISMSHEDRON AND A POLYTOPE DUAL TO IT

In Figure 7 a projection of the 3*6 - angular prismahedron with marked 18 vertices is shown.

It has three the 6 - angular prisms

$1)c_1c_2c_3c_4c_5c_6c_7c_8c_9c_{10}c_{11}c_{12}, 2)c_7c_8c_9c_{10}c_{11}c_{12}c_{13}c_{14}c_{15}c_{16}c_{17}c_{18},$

$1)c_1c_2c_3c_4c_5c_6c_{13}c_{14}c_{15}c_{16}c_{17}c_{18},$

and six the 3 - angular prisms

$4)c_1c_7c_6c_{13}c_{12}c_{18},$

*Figure 7.The 3*6 - angular prismahedron*

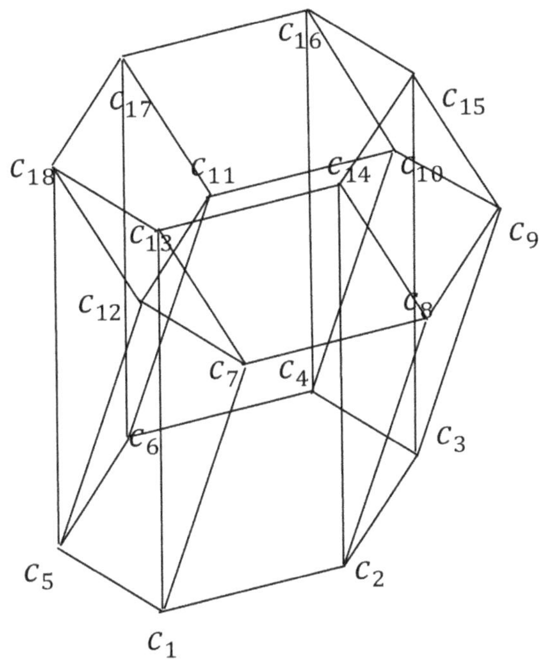

$5) c_1 c_2 c_7 c_{13} c_{14} c_8,$

$6) c_2 c_3 c_8 c_9 c_{14} c_{15}, 7) c_3 c_4 c_9 c_{10} c_{15} c_{16}, 8) c_4 c_5 c_{16} c_{10} c_{11} c_{17}, 9) c_5 c_6 c_{11} c_{12} c_{17} c_{18}.$

Analyzing contacts of the prisms on coinciding sets of symbols, one construct a projection of the polytope dual to the 6 * 3 - angular prismahedron (Figure 8).

In polytope in Figure 8 one can single out 18 tetrahedrons:

$T_1 = 1234, T_2 = 1235, T_3 = 1236,$

$T_4 = 1237, T_5 = 1238, T_6 = 1239, T_{10} = 3289, T_{11} = 1394, T_{12} = 1398,$
$T_7 = 1265, T_8 = 1267, T_9 = 2378,$

$T_{13} = 4325, T_{14} = 4315, T_{15} = 4215, T_{16} = 3267, T_{17} = 1378, T_{18} = 1278.$

*Figure 8. Polytope dual to the 3*6 - angular prismahedron*

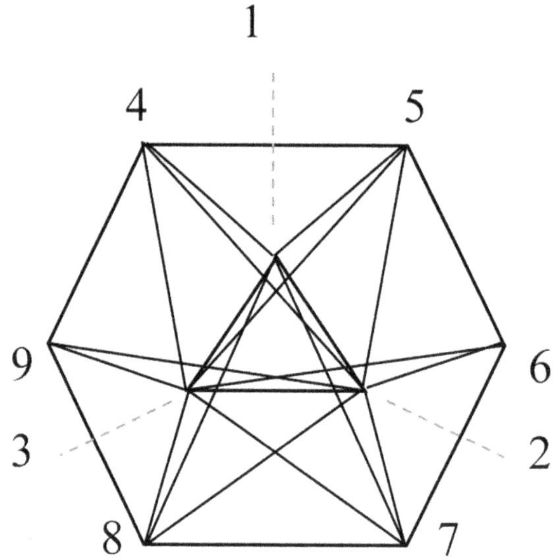

The polytope has 27 edges with different values of incidence to tetrahedrons: edge

$45 \in T_{13}, T_{14}, T_{15}; 12 \in T_1, T_8, T_{15}, T_{18}; 15 \in T_2, T_7, T_{14}, T_{15}; 14 \in T_1, T_{11}, T_{14}, T_{15}; 16 \in T_3, T_7, T_8;$

$56 \in T_7; 19 \in T_6, T_{11}, T_{12}; 94 \in T_{11}; 13 \in T_1, T_6, T_{11}, T_{12}, T_{14}, T_{17}; 42 \in T_1, T_{13}, T_{15}; 35 \in T_2, T_{13}, T_{14};$

$43 \in T_1, T_{11}, T_{13}, T_{14}; 32 \in T_1, T_6, T_9, T_{10}, T_{13}, T_{16}; 93 \in T_6, T_{10}, T_{12}; 26 \in T_3, T_7, T_8, T_{16}; 92 \in T_6, T_{10};$

$36 \in T_3, T_{16}; 17 \in T_4, T_8, T_{17}, T_{18}; 52 \in T_2, T_7, T_{13}, T_{14}; 87 \in T_9, T_{17}, T_{18}; 18 \in T_5, T_{12}, T_{17};$

$37 \in T_4, T_9, T_{16}, T_{17}; 82 \in T_5, T_9, T_{10}, T_{18}; 27 \in T_4, T_8, T_9, T_{16}, T_{18}; 83 \in T_5, T_9, T_{10}, T_{12}, T_{17};$

$76 \in T_8, T_{16}; 98 \in T_{10}, T_{12}.$

Thus, edges 56, 94 have incidence value equal to 1; edges 92, 36, 76 have incidence value equal to 2; edges 45, 16, 19, 42, 35, 87 have incidence value equal to 3; edges 15, 14, 43, 93, 26, 17, 52, 82 have incidence value equal to 4; edges 27, 83 have incidence value equal to 5; edges 12, 13, 32 have incidence value equal to 10.

In addition, can to note that vertices 1, 2, 3 are incident to 14 tetrahedrons; vertices 4, 5 are incident to 5 tetrahedrons; vertices 6, 9 are incident to 4 tetrahedrons; vertices 7, 8 are incident to 6 tetrahedrons.

This method can be used for construction of dual polytopes to many discussed in this paper, the polytopes with simultaneous use of two - dimensional faces of different shapes. All these polytopes are poly - incident, i.e. they have edges (and vertices) with different values of incidence.

THE *N* – PRISMAHEDRONS AND POLYTOPES DUAL TO THEM

Let's call the n - prismahedron a polytope of dimension 4, with facets in the form of prisms with n - angular bases. In item "The 4*3 – angular prismahedron and its dual polytope" a 3 - prismahedron was considered. Now let'us turn to the study of the n - prismahedrons for any n.

Theorem 2

The product of two the n - angular polygons is an n - prismahedron with the number of facets $2n$.

Theorem 2 follows immediately from theorem 1 when $m = t = n$.

Corollary. In a polytope dual to the n - prismahedron the number of vertices is equal to $2n$, the number of edges is equal to $2n + n^2$, the number of the two - dimensional faces is equal to $2\ n^2$, the number of the three - dimensional faces is equal to n^2. This follows directly from the corollary to theorem 1.

Theorem 3

Polytopes dual to the n - prismahedrons are those composed of the tetrahedrons.

For $n = 4$ the dual polytope is a 4 – cross - polytope, for $n = 10$ the dual polytope is a previously unknown polytope consisting of 100 tetrahedrons (100 - cell one) with a degree of incidence equal to 5, for other values of n

≥ 3 the dual are the poly - incident polytopes with a medium degree of edge incidence tending to 6 with an infinite increase of n.

Proof

As polytopes dual to the n - prismhedrons have the number of vertices equal to $2n$, owing to the inverse of inclusion relation, the number of edges is equal to $2n + n^2$, the number of 2 - faces is equal to $2n^2$, the number of 3 - faces is equal to n^2.

Then each 3 - figure has the number of 2 -faces $2\frac{2n^2}{n^2} = 4$ (multiplication by 2 is due to the fact that each 2-face belongs at once to two 3 -shapes). Thus, a 3 - figure is a tetrahedron, since only tetrahedron, as a 3 - figure, has 4 the 2 - faces. As everybody knows, the number of edges in a tetrahedron is 6. Can to write this number on the basis of set numbers of different dimension in a dual polytope: $6 = k\frac{2n + n^2}{n^2}$, k is the degree of edge incidence (the edge belongs to k-tetrahedrons) under the assumption that all the edges have this degree of incidence, the same for all. Hence, can to find the degree of incidence

$$k = \frac{6n}{n+2}, n \geq 3. \qquad (2)$$

Equation (2) in integers has two solutions: $k = 4$ when $n = 4$, and $k = 5$ when $n = 10$. For the other possible values of n, according to (2), the degree of incidence of edges is a fractional value. This means that in these cases in a dual polytope edges with varying degrees of incidence present simultaneously, i.e. it is a poly - incident polytope.

Figure 3, Figure 9, Figure 10 give examples of the 3 - prismahedron, the 5 - prismahedron, the 6 - prismahedron respectively.

The known hypercube is the 4 - prismahedron.

For $n = 4$ in a dual polytope the number of vertices is $2n = 8$, the number of edges is $2n + n^2 = 24$, the number of the 2 - faces is $2n^2 = 32$, the number of the 3 - faces is $n^2 = 16$. This is a 4 – cross - polytope dual to a hyper - cube. Each edge of this polytope has a degree of incidence 4.

Figure 9. The 5 - prismahedron

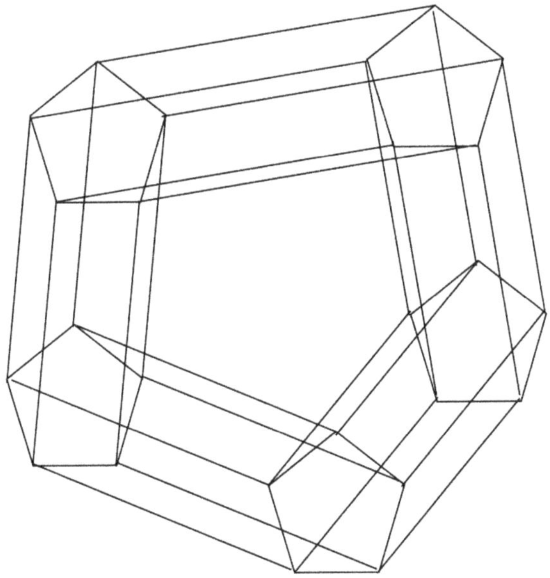

Figure 10. The 6 - prismahedron

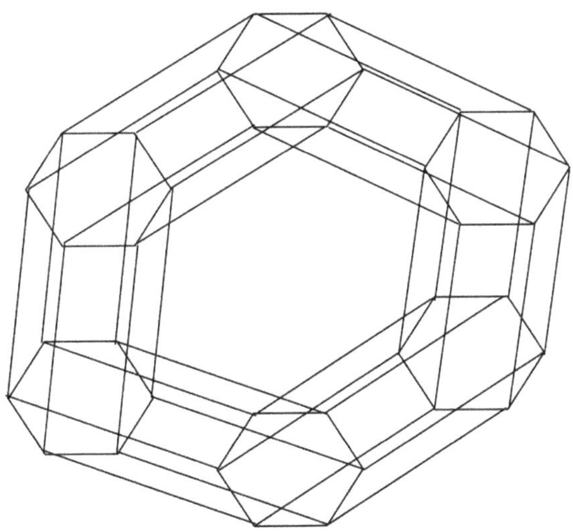

For $n = 5$ in a dual polytope the number of vertices is $2n = 10$, the number of edges is $2n + n^2 = 35$, the number of the 2 - faces is $2n^2 = 50$, the number of the 3 - faces is $n^2 = 25$. In the projection on the plane this polytope is presented in Figure 11.

In Figure 11, the tetrahedrons are: 1)*abgh,* 2)*dckh,* 3)*kcnd,* 4)*efnd,* 5) *agfe,* 6)*hkgn,* 7)*knfh,* 8)*fngk,* 9)*gfhn,* 10)*ghfk,* 11)*ahfg,* 12)*gbkh,* 13)*hcnk,* 14)*fkdn,* 15)*egnf,* 16)*abec,* 17)*bcad,* 18)*cdeb,* 19)*edac,* 20)*aebd,* 21)*agbe,* 22)*bhac,* 23)*kcbd,* 24)*ndec,* 25)*efad.* Every vertex of the polytope belongs to 10 tetrahedrons. Every 2 - face belongs to two tetrahedrons (also as for *n = 3; 4).* From 35 edges 10 edges are incident to 3 tetrahedrons: $gb \in$ 1), 12), 21); $bk \in$ 2), 12), 23);$hc \in$ 2), 13), 22); $nc \in$ 3), 13), 24); $kd \in$ 3), 14), 23); $fd \in$ 4), 14), 25);$en \in$ 4), 15), 24); $eg \in$ 5), 15), 21);$af \in$ 5), 11), 25); $ah \in$ 1), 11), 22).

From 35 edges following 15 edges are incident to 4 tetrahedrons:

$ag \in$ 1), 5), 11), 21); $bh \in$ 1), 2), 12), 22); $kc \in$ 2), 3), 13), 23); $nd \in$ 3), 4), 14), 24);

Figure 11. Polytope dual to the 5 - prismahedron

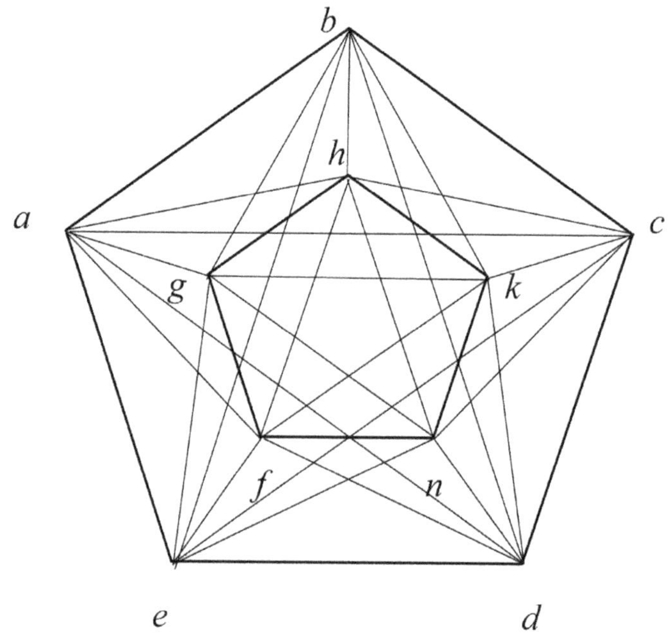

ef ∈ 4), 5), 15), 25); *eb* ∈ 16), 18), 20), 21); *bd* ∈ 17), 18), 20), 23); *ac* ∈ 16), 17), 19), 22); *ec* ∈ 16), 18), 19), 24); *ad* ∈ 17), 19), 20), 25); *hf* ∈ 7), 9), 10), 11); *hn* ∈ 6), 7), 9), 13); *kg* ∈ 6), 8), 10), 12); *kf* ∈ 7), 8), 10), 14); *gn* ∈ 6), 8), 9), 15).

From 35 edges following 10 edges are incident to 6 tetrahedrons:

ab ∈ 1), 16), 17), 20), 21), 22); *bc* ∈ 2), 16), 17), 18), 22),23); *cd* ∈ 3), 17), 18), 19), 23), 24);

ed ∈ 4), 8), 19), 20), 24), 25); *ae* ∈ 5), 16), 19), 20), 21), 25); *gh* ∈ 1), 6), 9), 10), 11), 12); *hk* ∈ 2), 6), 7), 10), 12), 13); *kn* ∈ 3), 6), 7), 8), 13), 14); *fn* ∈ 4), 7), 8), 13), 14); *gf* ∈ 5), 8), 9), 10), 11), 15). Thus, the average value of incidence degree is equal to: $\bar{k} = \frac{10 \cdot 6 + 15 \cdot 4 + 10 \cdot 3}{35} = \frac{30}{7}$ and coincides exactly with the degree of incidence according to equation (2).

By increasing *n*, the structure of a dual polytope gets greatly more complicated and there are difficulties in its constructing. For $n = 6$ in a dual polytope the number of vertices is $2n = 12$, the number of edges is $2n + n^2 = 48$, the number of the 2 - faces is $2n^2 = 72$, the number of the 3 - faces is $n^2 = 36$. This polytope is presented in Figure 12 in the projection on the plane.

In Figure 12 tetrahedrons are:

hmng, 2) *acdf,* 3) *hkmg,* 4) *abcf,* 5) *hkmn,* 6) *abcd,* 7) *hkno,* 8) *abde,* 9) *gmno,* 10) *fcde,* 11) *kmno,* 12) *bcde,* 13) *gkmo,* 14) *fbce,* 15) *ghko,* 16) *fabe,* 17) *ghno,* 18) *fade,* 19) *hkmb,* 20) *abck,* 21) *kmnc,* 22) *bdcm,* 23) *mnod,* 24) *cden,* 25) *gone,* 26) *fedo,* 27) *hgof,* 28) *afeg,* 29) *ghka,* 30) *fabh,* 31) *kmbc,* 32) *mncd,* 33) *oned,* 34) *gofe,* 35) *hgaf,* 36) *hkab.* 1)

From 48 edges each of 12 edges is incident to 7 tetrahedrons from the following list:

ab, bc, cd, de, fe, af, hk, km, mn, no, og, hg; each of 24 edges is incident to 4 tetrahedrons: *ko, gm,hn, ho, kg, mo, kb, oe, fc, be, ad, ac, fd, bd, mc, gf, hm, gn, kn, ae, bf, ce, nd, ha;* each of 12 edges is incident to 3 tetrahedrons: *bm, kc, ge, fo, md, cn, ga, hf, ne,od, hb, ak.*

Figure 12. Polytope dual to the 6 - prismahedron

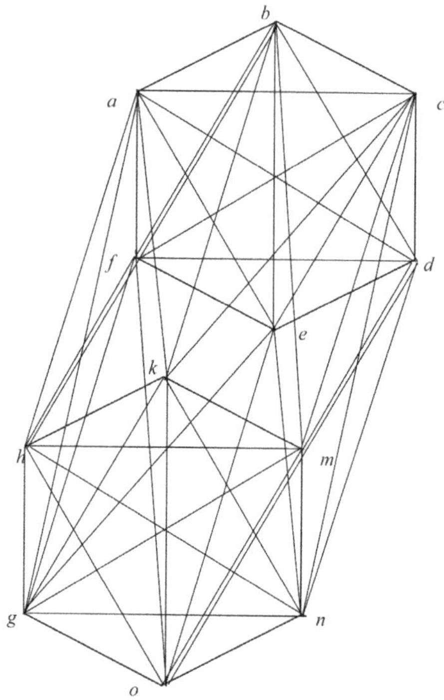

Thus, the average value of edges incidence degree is equal to $\bar{k} = \dfrac{12 \cdot 7 + 24 \cdot 4 + 3 \cdot 12}{48} = 4,5$, which fits exactly with the degree of incidence k according to equation (2).

Every vertex of this polytope is incident to 12 tetrahedrons. From 72 triangles, which are faces of tetrahedrons, some have degree of incidence 3 and the rest have degree of incidence 2.

The construction of a dual polytope with $n = 10$ is particularly interesting, since according to equation (2) in this case the degree of incidence of edges is the same for the whole polytope and equals to 5. For $n = 10$ in a dual polytope, the number of vertices is $2n = 20$, the number of edges is $2n + n^2 = 120$, the number of the 2 - faces is $2n^2 = 200$, the number of the 3 - faces is $n^2 = 100$. This polytope is presented in Figure 13 in the projection on the plane.

Tetrahedrons in Figure 13 are:

1) *dkhc,* 2) *cgbh,* 3) *fagb,* 4) *aefm,* 5) *demk,*

Figure 13. Poytope dual to the 10 - prismahedron

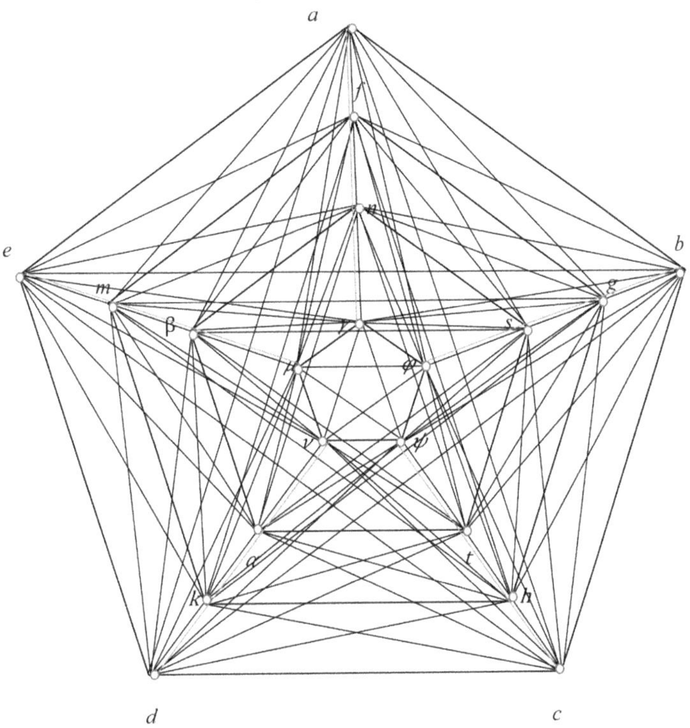

6) *kthα,* 7) t*hsg,* 8) n*fsg,* 9) m*fnβ, 1*0) *β*m*kα,*

11) *α*t*ψν, 12) φψ*st, *13) γφn*s, 14*) μγnβ,* 15) *βμαν,*

16) *φψγμ, 17) φψμγ, 18) φψγν,* 19) *γψνφ, 20) φγνμ,*

21) *d*hbc, *22) cgab,* 23) *bfae, 24)* daem, *25)* ekcd,

26) ktg*h, 27*) hsf*g, 28*) mgn*f, 29*) fkm*β, 30)* mhk*α,*

31) dc*νψ,* 32) *φ*c*bψ, 33*) a*γ*b*φ, 34) μ*ae*γ*, *35) ν*de*μ*,

36) $vd\psi\mu$, 37) $cv\varphi\psi$, 38) $v\varphi\psi b$, 39) $a\varphi\gamma\mu$, 40) $e\gamma v\mu$,

41) $\varphi\gamma v\mu$, 42) $\varphi vt\psi$, 43) $\varphi\gamma\psi s$, 44) $\mu n\varphi\gamma$, 45) $\beta\gamma v\mu$,

46) kdth, 47) hcsg, 48) gbnf, 49) afmβ, 50) αemk,

51) αkhc, 52) thgb, 53) asgf, 54) fnem, 55) mdkβ,

56) $k\alpha t\psi$, 57) thφs, 58) sgγn, 59) nfμ β, 60) αvmβ,

61) thαv, 62) ψsgt, 63) snfφ, 64) n$\beta\gamma\mu$, 65) k$\beta\mu\alpha$,

66) kt$\alpha\beta$, 67) thsα, 68) sgnt, 69) fnmg, 70) mn$\beta\alpha$,

71) t$\alpha v\beta$, 72) ts$\alpha\psi$, 73) sntφ, 74) ns$\beta\gamma$, 75) $\beta\alpha\mu$n,

76) mhαv, 77) tkgψ, 78) sfhφ, 79) γnmg, 80) kf $\beta\mu$,

81) dec$v$, 82) cdbψ, 83) tacφ, 84) ebγa, 85) daμe,

86) dmh$v$, 87) ckgψ, 88) bfhφ, 89) amgγ, 90) efkμ,

91) kc$\alpha\psi$, 92) thbφ, 93) sgγa, 94) enfμ, 95) m$v\beta$d,

96) kμeα, 97) ma$\beta\gamma$, 98) fnbφ, 99) fgψc, 100) thd$v$.

In the given above list of the tetrahedrons each line corresponds to a specific type of the tetrahedron projection, which differ only by angle of rotation relative to the polytope projection center divisible by 72 degrees. The structure of a 100 - cell polytope is quite complicated, but it is defined by enumerating all the tetrahedrons. If all the edges of a polytope have the same degree of incidence, the degree of incidence of vertices and triangles, which are faces of tetrahedrons, is variable.

By increasing the value of n in equation (2), the average degree of edges incidence tends to its maximum value of 6.

Theorem 3 is proved.

Note. It should be noted that the number of elements of different dimension of the polytope in Figure 7 is 6 times smaller than the corresponding values of a regular polytope consisting of 600 tetrahedrons (600 - cell polytope, Coxeter, 1963) with the same degree of edges incidence. Since the proofs for the existence of 600 - cell polytope in the works of Stringham (Stringham, 1880) and Coxeter (Coxeter, 1963) contain incorrectness (Zhizhin, 2014), there is no specific enumeration of tetrahedrons and their images do not carry specific information, the structure of the proposed 600 - cell polytope remains unknown.

POLYTOPES DUAL TO THE POLYTOPIC PRISMAHEDRONS OF HIGH DIMENSION

The most effective for describing the nature of the structure of high - dimensional polytopic prismahedrons are their dual polytopes. In Chapter 4, for example, an analytical form and an image of the structure of the product of the tetrahedron per square (equation (25) and Figure 13 in Chapter 4) were presented. This polytopic prismahedron has dimension equal 5. Figure 13 Figure 13 is a bit complicated. Let us construct a polytope dual to the polytope in Figure 13. According to the definition of a dual polytope, it is necessary to select the facets of a given polytope and connect the centers of those facets that share common three - dimensional faces. It is easy to see that these three - dimensional faces are tetrahedrons, and the facets are polytopes formed by two tetrahedrons connected by edges on the same vertices (Figure 14). It can be shown that facets have dimension 4.

Figure 14. Facets of product of the tetrahedron by square

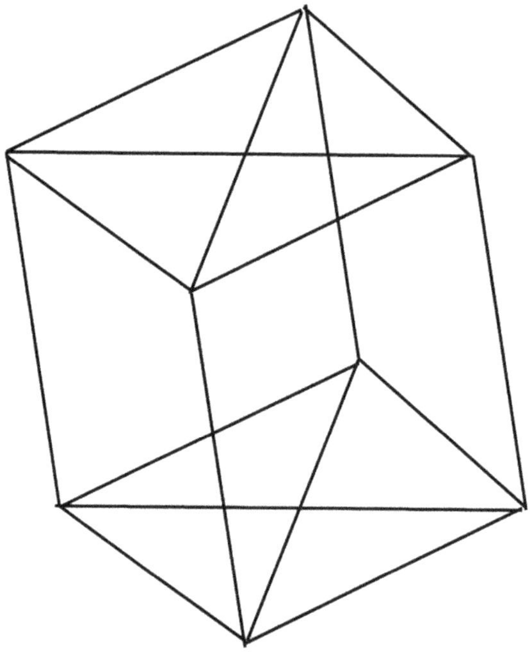

Indeed, this polytope has 8 vertices $\left(f_0 = 8\right)$, 16 edges $\left(f_1 = 16\right)$, 14 flat faces $\left(f_2 = 14\right)$, 6 three - dimension faces $\left(f_3 = 6\right)$. Substituting these values into the Euler - Poincaré equation (2) of Chapter 2, can to find that it is satisfied at a value of n equal to 4. Therefore, the dimension of the facet is 4. The facet in Figure 14 is 4. If, by definition, one designate each facet as a vertex, then can to get a closed chain of polytopes of dimension 4 connected to each other along a common tetrahedron (Figure 15).

Thus, the polytope dual to polytope on Figure 13 Chapter 4 is square in vertices one are located the polytopes with dimension 4 and edges correspondent the common three - dimension figures - tetrahedrons.

Consider the polytopic prismahedron in Figure 12 of Chapter 4. This is the product of the 4 – cross - polytope by triangle. The polytope has dimension 6. One construct a dual polytope. For this purpose, can to select five - dimensional facets in its boundary complex. Such polytopes are products of a 4 – cross - polytope by one - dimensional segment. There are three such polytopic prismahedrons in the boundary complex in Figure 12 of Chapter 4. Each of them has two the 4 –cross - polytopes and polytopes of dimension 5

Figure 15. Polytope dual to product square by tetrahedron

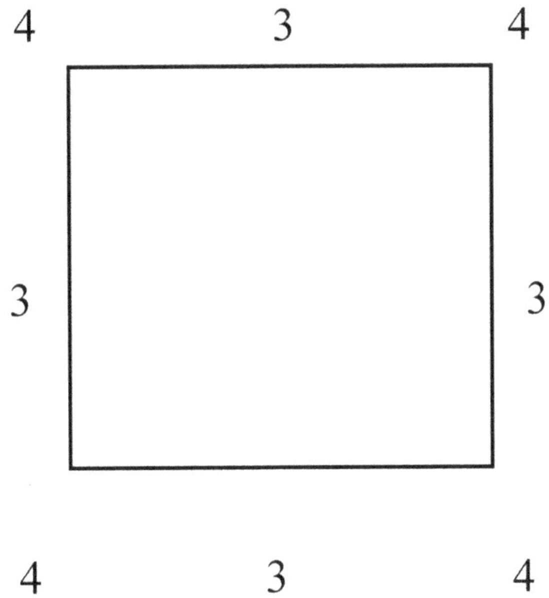

intersect in each of these polytopes. A closed chain of polytopes of dimension 5 is formed (Figure 16).

Therefore, the polytope dual to the plate in Figure 12 of Chapter 4 is a triangle with vertices in the form of polytopes of dimension 5 and edges in the form of polytopes of dimension 4.

Thus, dual polytopes distinguish their structure from polytopes of large dimension in polytopic prismahedrons, one can say their factor - structure is distinguished.

CONCLUSION

It has been established, that the polytopic prismahedrons studied in Chapter 4 have, in the general case, another surprising property: polytopes dual to the polytopic prismahedrons have faces (vertices, edges) with different incidence values with respect to higher dimensional faces in this polytope. Such polytopes form a new class: a class of poly - incident polytopes, which are obviously incorrect, but have certain symmetry elements. Analytical analysis of various types of polytopes, dual to the polytopic prismahedrons, is carried out. Images

Figure 16. The polytope dual to product the 4 – cross - polytope by triangle

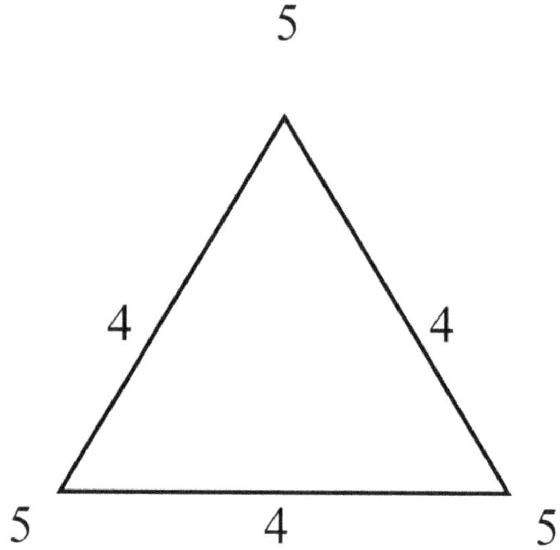

of these polytopes are constructed. It is established, that a polytope dual to the product of two the n - angular polygons (n -prismahedron), with n being equal to 10, is an earlier unknown polytope from 100 tetrahedrons. All tetrahedrons entering this polytope are listed, its image is given. It is established that for n = 5 and n = 10 polytopes, dual polytope to the n - prismehedrons have edges with constant incidence values (4 and 5, respectively). For other values of n, the dual polytope to the n - prismahedrons are poly-incident polytopes. For the polytopic prismahedrons with a dimension greater than four, the concept of a factor - structure is introduced, the vertices and edges of which are polytopes of higher dimension. The factor - structure is a polytope dual to a certain polytopic prismahedron.

For the first time, analytical laws for the dependence of the incidence coefficients of the elements of the n - cubes, the n - simplexes, and the n – cross - polytopes on the dimension of the polytope are established. According to these laws, the incidence coefficients increase linearly with increasing n, especially this increase occurs in the n – cross - polytopes.

REFERENCES

Bogdanov, A. A. (1989). Tectology. General Organizational Science. Moscow: *E*conomics.

Buckour, I., & Delanu, A. (1972). Introducti*on to the theory of categories. Moscow:* World.

Coxeter, H.S.M. (1963). Regular Polytopes. New *York, London: John Wile & Sons, INC.*

Feyman, R. (1968). The nature of physica*l laws. Moscow:* World.

Grunbaum, B. (1967). Convex Polytopes. London: Springer.

Hegel, G. V. (1998). The Science of Logic. Moscow: Th*ought.*

Poincare, A. (1895). Analysis situs. J. de é Ecole Poly*technique, 1, 1* – 121.

Stringham, W. I. (1880). Regular figures in n - *dimensional space. American Jo*urnal of Mathematics, 3(1), 1–14. doi:10.2307/2369441

Whitehead, A. N. (1990). Selected works on philosophy. Moscow: Progress.

Zhizhin, G. V. (2014*). World – 4D. St. Petersburg:* Polytechnic Service.

Zhizhin, G.V. (2016). N – prismah*edrons and* their dual polytopes. International Journal Chemical Modeling, 5(4).

Zhizhin, G. V. (2017). Chemical *Compound Structures and the Higher Dimensi*on of Molecules: emerging Re*search and Opportunities. Hershey, PA: IGI Global.*

Ziegler, G. (1995). *Lectures on polytopes. New Yo*rk: Springer. doi:10.1007/978-1-4613-8431-1

KEY TERMS AND DIFINITIONS

Duality in Polytopes: *If* the facets of one polytope correspond to the vertices of another polytope, and the edges of one polytope correspond to common elements of the facets of another polytope, then these polytopes are dual.

Incidence in Polytopes: Incidence in polytopes define the number of elements of higher dimension the given element of lower dimension belongs.

Poly - Incident Polytopes: Polytopes in which elements of lower dimension have different incidence values for elements of higher dimension. Polytopes that are dual to polytopes products are poly - incident polytopes.

Polytope with Factor – Structure: Polytope in which vertices are polytopes with same dimension k, and edges are polytopes with dimension k – 1.

Chapter 6
The Detailed Structure of *n* – Cross – Polytopes and Polytopes With Their Participation

ABSTRACT

The structure of the n – cross - polytopes for large values of n with an exact enumeration of elements of various dimensions entering their boundary complexes is studied in detail. Examples of chemical compounds with the structure of the n – cross - polytopes are given. It is shown that the polytopic prismahedrons can be polytopes, which simultaneously include the n - simplexes and the n – cross- polytopes as faces. It is established that the list of the n - simplexes and the n – cross - polytopes of different dimensions represented by Gosset as faces of this polytope contradicts the analytic laws of increasing incidence coefficients with increasing n, first established in Chapters 1, 3, 4 of this book. This testifies to the impossibility of the existence of Gosset's polytope.

INTRODUCTION

In 1900, Gosset (who did not have an education in mathematics) announced (but did not prove) the existence of a polytope of dimensionality 8 (Gosset, 1900), in which both the cross - polytopes and the simplexes of dimension

DOI: 10.4018/978-1-5225-6968-8.ch006

7 are simultaneously facets. He listed the faces of this polytope and the smaller dimension. This work it was supported by Coxeter (Coxeter, 1963). Later, various images of Gosset's polytope were obtained. However, these visualizations were more of an artistic nature. According to these images, it is impossible accurately to list the faces specified by Gosset included in the polytope. In this chapter there is consider in detail the structure of the n – cross - polytopes for large values of n. The question of the possible simultaneous presence of high dimensionality in polytopes as faces and the cross - polytopes and the simplexes is analyzed. In particular, the question of the possibility of the existence of a polytope with the participation of faces is considered according to the list indicated by Gosset.

THE *N* – CROSS - POLYTOPE IS SIMPLICIAL POLYTOPE

In Chapter 2 it was noted that the n - cross – polytope is simplicial polytope (Ziegler, 1995; Grunboum, 1967). For $n = 4$ this is clear, so the facets of the 4 – cross - polytope are the tetrahedrons which are simplexes and all faces of the tetrahedron are simplexes. For $n > 4$ this statement is not evident. Let is consider the 5 – cross – polytope for example. The image of this polytope it was presented in Chapter 2 (Figure 10). According to equation (6) of Chapter 2 for the 5 – cross - polytope there are

$$f_0(5) = 10, f_1(5) = 40, f_2(5) = 80, f_3(5) = 80, f_4(5) = 32.$$

Thus, the 5 – cross – polytope has 10 vertices, 40 edges, 80 flat faces (triangles), 80 tetrahedrons and 32 polytopes with dimension 4. However, from equation (6) of Chapter 2 no follow which form of the polytopes with dimension 4. One listed this elements with dimension 0, 1, 2, 3, used Figure 10 of Chapter 2. The vertices are 0, 1, 2, 3, 4, 5, 6, 7, 8, 9. The edges are 10, 19, 18, 17, 15, 14, 13, 12, 20, 29, 28, 26, 25, 24, 23, 30, 39, 37, 35, 34, 40, 48, 47, 46, 45, 59, 58, 57, 56, 60, 69, 68, 67, 70, 79, 78, 80, 89, 90.

The triangles are

109, 108, 107, 104, 103, 102, 198, 197, 195, 193, 192, 187, 185, 184, 182, 175, 174, 173, 154, 153, 152, 143, 142, 132, 209, 208, 206, 204, 203, 298, 296, 295, 293, 286, 285, 284, 265, 264, 263, 254, 253, 243, 309, 307, 306, 304, 397, 396, 395, 376, 375, 374, 365, 364, 354, 408, 407, 406, 487, 485,

476, 465, 486, 475, 598, 597, 596, 587, 586, 576, 609, 608, 607, 698, 697, 687, 709, 708, 798, 809.

The tetrahedrons with listed triangle faces are

1098, 1097, 1093, 1092, 1087, 1084, 1082, 1074, 1073, 1043, 1042, 1032, 1987, 1985,

1982, 1975, 1974, 1973, 1953, 1952, 1932, 1875, 1874, 1854, 1852, 1842, 1754, 1753,

1743, 1543, 1542, 1532, 1432, 2098, 2096, 2093, 2086, 2084, 2064, 2063, 2043, 2986,

2985, 2965, 2963, 2953, 2865, 2864, 2854, 2654, 2653, 2643, 2543, 3097, 3096, 3076,

3074, 3064, 3976, 3975, 3965, 3765, 3764, 3754, 3654, 3654, 4087, 4086, 4076, 4876,

4875, 4765, 4865, 5987, 5986, 5976, 6098, 6097, 6087, 6987, 7098.

The number of vertices, edges, triangles and tetrahedrons defined according to Figure 10 of Chapter 2 exactly coincide with the number of elements of dimension zero, one, two and three, calculated according to formula (6) of Chapter 2. The number of these figures is limited by the condition that there are no edges connecting the vertices opposite in Figure 10.

In Figure 10 of Chapter 2, it is possible to single out the projections of polytopes the 4 - simplex. One list these projections

$$s_1 = 12345, s_2 = 23456, s_3 = 34567, s_4 = 45678, s_5 = 56789, s_6 = 67890, s_7 = 78901,$$
$$s_8 = 89012, s_9 = 90123, s_{10} = 01234, s_{11} = 13579, s_{12} = 24680, s_{13} = 13457,$$
$$s_{14} = 24568, s_{15} = 35679, s_{16} = 46780, s_{17} = 57891, s_{18} = 68902, s_{19} = 79013,$$
$$s_{20} = 80124, s_{21} = 91235, s_{22} = 02346, s_{23} = 34760, s_{24} = 12458, s_{25} = 92356, s_{26} = 45871,$$
$$s_{27} = 56982, s_{28} = 67903, s_{29} = 78014, s_{30} = 89125, s_{31} = 09236, s_{32} = 01347.$$

There are no other polytopes of the 4 - simplex in Figure 10 of Chapter 2. The analysis of the listed facets of the 5 - cross - polytope shows that each vertex of the 5 – cross - polytope belongs to 16 facets (type 4 - simplex). For example, the vertex

$1 \in s_1, s_7, s_8, s_9, s_{10}, s_{11}, s_{13}, s_{17}, s_{19}, s_{20}, s_{21}, s_{24}, s_{26}, s_{29}, s_{30}, s_{32}.$

Each edge belongs to 8 polytopes the 4 - simplex. For example, the edge

$12 \in s_1, s_8, s_9, s_{10}, s_{20}, s_{21}, s_{24}, s_{30}.$

Each triangle belongs to 4 polytopes the 4 – simplex. For example, triangle $123 \in s_1, s_9, s_{10}, s_{21}$. Consequently, there are 12 tetrahedrons adjacent to each tetrahedron on the whole sides. Each tetrahedron belongs to 2 polytopes the 4 – simplex. For example, tetrahedron $1234 \in s_1, s_{10}$. Thus, the incident coefficients of elements of the 5 – cross – polytope to relation their facets have senses (see Chapter 5) $k_{b(0),b(4)} = 16, k_{b(1),b(4)} = 8, k_{b(2),b(4)} = 4, k_{b(3),b(4)} = 2.$

In Figure 10 of Chapter 2, 15 polytopes of the 4 – cross - polytopes can also be identified. For example, five such polytopes can be obtained by removing one of the pairs of oppositely located vertices 02345789, 10345689, 01245679, 01235678, 12346789. However, the total number of these polytopes 15 is less than the required number of polytopes of dimension 4, defined by the formula (6) of Chapter 2, equal to 32 $(f_4(5) = 32)$. The combination of facets of different shapes in one polytope, even if it were possible, violates the condition of the correctness of the polytope. At the same time, the number of facets determined in Figure 10 of Chapter 2 in the form of a 4 -simplex is exactly equal to the number of facets by the equation (6) of Chapter 2, i.e. it is 32. Substituting the values, obtained from Figure 10 of Chapter 2 for the numbers of elements of different dimensions, into the Euler - Poincaré equation (2) of Chapter 2 (Poincare, 1895), can to find

$10 - 40 + 80 - 80 + 32 = 2.$

Hence, the Euler - Poincaré equation holds and the 5 – cross - polytope is a simplicial polytope.

The projection of a 6 – cross - polytope onto a two - dimensional plane is shown in Figure 11 of Chapter 2. According to equation (6) of Chapter 2, this polytope has the following values of elements of different dimensions

$$f_0(6) = 12, f_1(6) = 60, f_2(6) = 160, f_3(6) = 240, f_4(6) = 129, f_5(6) = 64.$$

The 6 – cross – polytope has six pairs opposite vertices not connected by edge (they are arranged vertically)

$$\begin{matrix} a_1 & a_2 & a_3 & a_4 & a_5 & a_6 \\ a_7 & a_8 & a_9 & a_{10} & a_{11} & a_{12} \end{matrix} \tag{1}$$

There are also as in the 5 – cross – polytope facets are simplexes, in this case the simplexes have dimension 5. The incident coefficients of elements of the 6 – cross – polytope to relation their facets have senses

$$k_{b(0),b(5)} = 32, k_{b(1),b(5)} = 16, k_{b(2),b(5)} = 8, k_{b(3),b(5)} = 4, k_{b(4),b(5)} = 2.$$

The equation (7) of Chapter 2 determines the location of the elements of dimension i around elements of dimension $i-1$. For example, each triangular face in a 6 – cross - polytope is common to all four neighboring tetrahedrons. Each tetrahedron belongs to four the 4 - simplexes, and so on. Summarizing the results of the investigation of the cross-polytopes of dimensions 4, 5, and 6, one can to write a general expression for the incidence coefficient with respect to the facets of elements of the n – cross - polytopes of different dimensions

$$k_{b(n-1-j),b(n-1)} = 2j \tag{2}$$

where $n - 1 - j$ is the dimension of element of the n – cross – polytope.

The obtained analytical expressions enable us to construct an n – cross - polytope for any n. For example, the image of the 8 – cross - polytope is shown in Figure 1.

This polytope has 16 vertices, eight oppositely located pairs of which do not have an edge joint

Figure 1. The 8 – cross - polytope

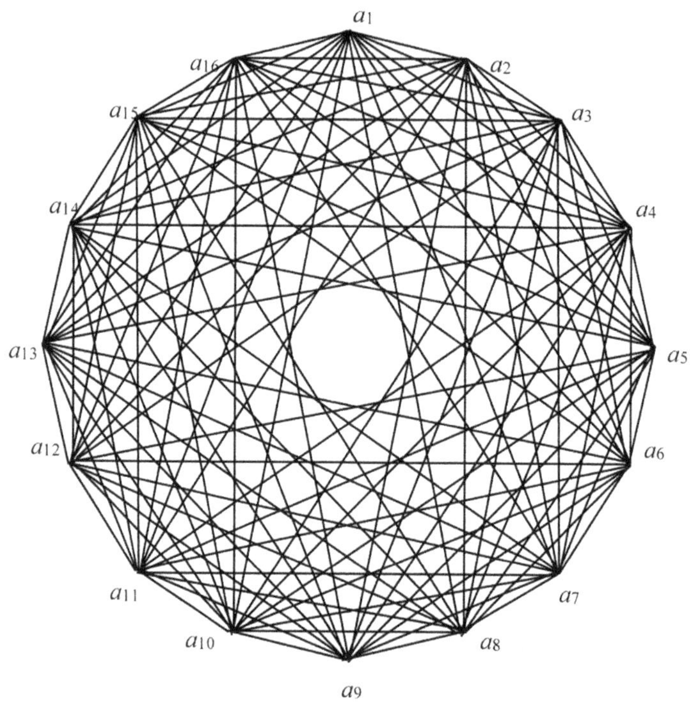

$$\begin{matrix} a_1 & a_2 & a_3 & a_4 & a_5 & a_6 & a_7 & a_8 \\ a_9 & a_{10} & a_{11} & a_{12} & a_{13} & a_{14} & a_{15} & a_{16} \end{matrix} \qquad (3)$$

The number of elements of different dimensions of the 8 – cross - polytope is determined by the series (formula (6) of Chapter 2)

$$f_0(8) = 16, f_1(8) = 112, f_2(8) = 448, f_3(8) = 160, f_4(8) = 1792, f_5(8) = 1792,$$
$$f_6(8) = 1024, f_7(8) = 256.$$

All the faces of the 8 - cross - polytope are simplexes, that is, this polytope is simplicial. The arrangement of faces of different dimensions with respect to each other is determined by the equality (7) of Chapter 2 and the series

$k_{b(0),b(8)} = 256, k_{b(1),b(8)} = 128, k_{b(2),b(8)} = 64, k_{b(3),b(8)} = 32, k_{b(4),b(8)} = 16, k_{b(5),b(8)} = 8,$
$k_{b(6),b(8)} = 4, k_{b(7),b(8)} = 2.$

THE POLYTOPES DUAL TO *N* – CROSS – POLYTOPES

It is known that a polytope dual to a 4 – cross - polytope is a polytope of 4 – cube (Ziegler, 1995). Let us find out how this condition of duality is realized specifically for a cross - polytope dimension greater than 4. Let the cross - polytope have dimension 5 (Figure 10 of Chapter 2). In this chapter, all 32 facets of this polytope, having the form of the polytope 4 - simplex, have already been listed. Each of these facets has a common tetrahedron with some other facet of this list. Let's imagine the connection structure of these facets in the form of a polytope whose vertices will correspond to the facets, and the edges will correspond to the common between the pair of facet tetrahedrons. By being, one will build a polytope dual to the 5 – cross - polytope. This will lead us to the symmetric structure in Figure 2.

One can to regard this structure as the factor - structure of the boundary complex of the 5 – cross - polytope, since each vertex of this structure is a polytope of dimension 4. In this structure, some edges overlap one another and two vertices in the center coincide. In order for the structure in Figure 2 to be more understandable, one slightly expand the coinciding edges and vertices (Figure 3).

It follows from Figure 3 that the obtained structure is the image of the projection of the polytope of the 5 - cube, since it contains images of 40 cubes adjacent to each other along entire sides. The total number of vertices (32), edges (80), squares (80), cubes (40), and 4 - cubes (10) in Figure 3 and Figure 2 exactly coincide with the required quantities of elements of different dimensions for the polytope 5 - cube.

Thus, Figure 2 is a new, previously unknown symmetrical image of the polytope 5 - cube. The previous image of the polytope 5 - cube (Zhizhin, 2014a) was a continuation of the image of the polytope 4 - cube (Stringham, 1880) into five - dimensional space. Obviously, similar images of n - dimensional cubes can be obtained even if $n > 5$.

Figure 2. The polytope dual to the 5 – cross - polytope

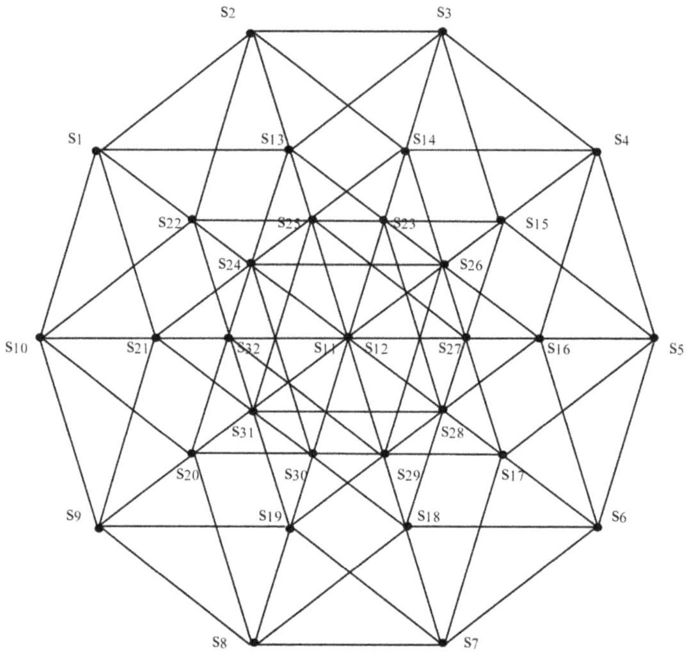

THE CROSS – POLYTOPES IN BOUNDARY COMPLEXES ANOTHER POLYTOPES

The absence of faces in the cross - polytopes in the form of the cross - polytopes does not mean that cross - polytopes can not be the faces of any polytopes. The cross - polytopes as faces always exist in polytopes, which are the product of a cross - polytope on some other polytopes (Chapter 4). Consider, for example, the product of a 4 – cross - polytope on a 4 - simplex. Their product will be a polytope of dimension 8, in which, by construction, there must necessarily be faces in the form of the 4 – cross - polytopes and the 4 - simplexes. The projection of this polytope on a two - dimensional plane is shown in Figure 4.

The facets of this polytope are the products of the 4 - cross - polytope on the tetrahedron (Figure 5) and the product of the 4 - simplex on the tetrahedron (Figure 6).

Figure 3. The polytope dual to the 5 – cross – polytope with slightly expand the coinciding edges and vertices

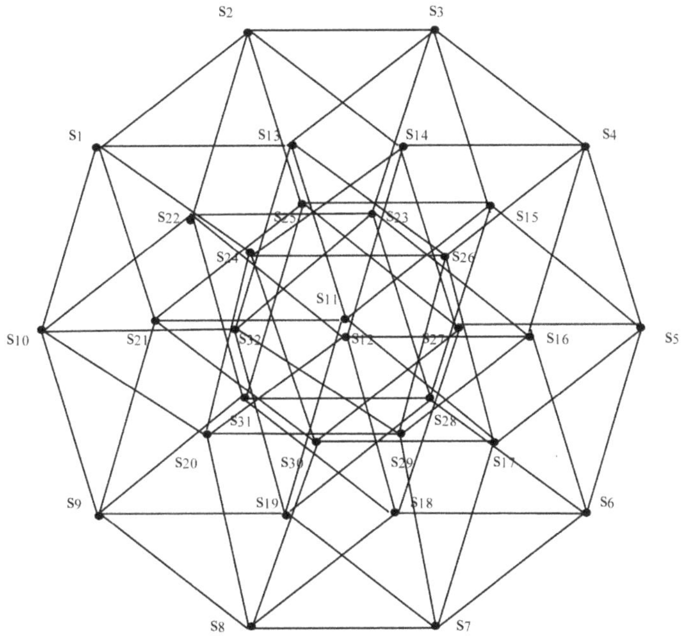

Figure 4. The product of the 4 – cross – polytope on the 4 - simplex

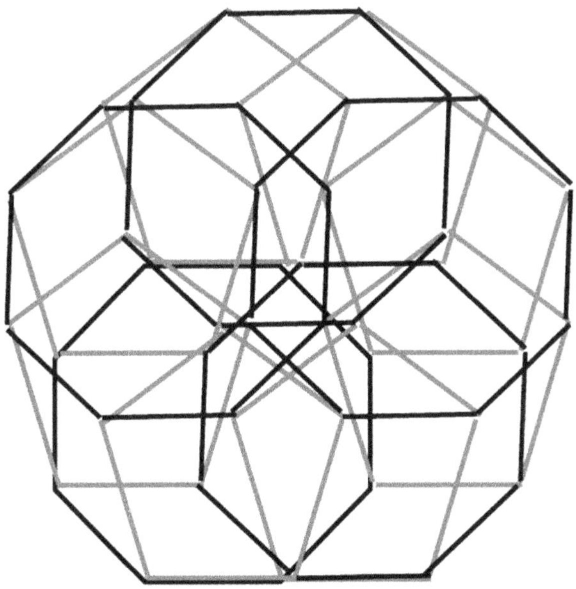

Figure 5. The product of the 4 – cross – polytope on the tetrahedron

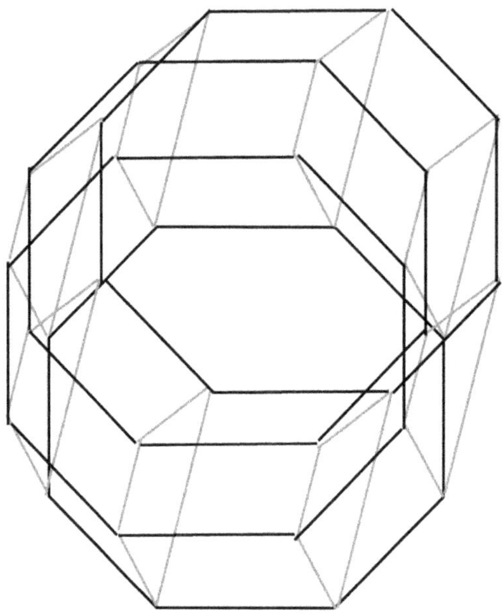

Figure 6. The product of the 4 – simplex on the tetrahedron

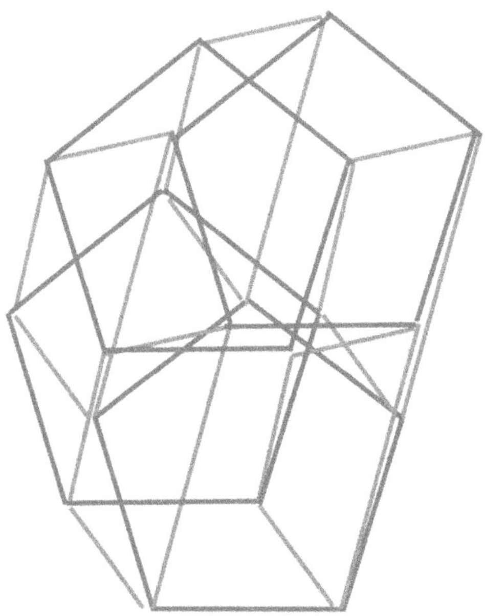

Since the 4 – cross - polytope has 8 vertices, and the 4 - simplex has 5 vertices, their product has 40 vertices $\left(f_0(8)=40\right)$. The number of facets in the form of the product of a 4 - cross - polytope to a tetrahedron is 5, and the number of facets in the form of a product of a 4 - simplex by a tetrahedron is 16. Therefore, the total number of facets is 21 $\left(f_7(8)=21\right)$.

To facilitate visualization, in Figure 4, black octagons represent the projections of the outer contours of the 4 - cross - polytopes onto a two - dimensional plane. Red pentagons represent the projections of the outer contours of a 4 - simplex onto a two - dimensional plane. In Figure 5, also, the black octagons represent the outer contours of the 4 – cross - polytopes projections onto the two-dimensional plane, and the blue quadrilaterals represent the outer contours of the tetrahedron projections, and the red pentagons represent the outer contours of the 4 - simplex projections.

Elements of dimension 6 in the product of the 4 – cross - polytopes by a 4 - simplex are the following figures:

1. The products of two - dimensional elements (triangles) of the 4 – cross - polytope by the 4 -simplex, i.e. 32 the polytopic prizmahedrons;
2. The products of two - dimensional elements (triangles) of the 4 - simplex by the 4 - cross - polytopes, i.e. 10 the polytopic prismahedrons;
3. The products of tetrahedrons of the 4 - simplex on tetrahedrons of the 4 – cross - polytopes, i.e. 80 the polytopic prismahedrons.

Thus, the total number of elements of dimension 6 in the product of polytopes under consideration is 122 ($f_6(8)=122$).

Elements of dimension 5 in the product of a 4 – cross - polytope to a 4 - simplex are the following figures:

1. The products of the 4 - cross polytope on the edges of the 4 - simplex, i.e. 10 the polytopic prismahedrons;
2. The products of the 4 - simplex on the edges of the 4 – cross - polytope, i.e. 24 the polytopic prismahedrons;
3. The products of plane elements (triangles) of the 4 – cross - polytopes onto tetrahedrons of the 4 - simplex, i.e. 160 items;
4. The products of plane elements (triangles) of the 4 - simplex onto tetrahedrons of the 4 - cross - polytopes, i.e. 160 items.

Thus, the total number of elements of dimension 5 in the product of polytopes under consideration is 354 ($f_5(8) = 354$).

Elements of dimension 4 in the product of a 4 – cross - polytope to a 4 - simplex are the following figures:

1. The product of vertices of the 4 – simplex onto the 4 – cross – polytope, i.e. 5 items;
2. The product of vertices of the 4 – cross – polytope onto the 4 – simplex, i.e. 8 items;
3. The product of flat elements of the 4 – cross – polytope onto a flat elements of the 4 – simplex, i.e. 320 elements;
4. The product of tetrahedrons of the 4 – cross – polytope onto edges of the 4 – simplex, i.e. 160 elements;
5. The product of tetrahedrons of the 4 –simplex onto edges of the 4 – cross - polytope, i.e. 120 elements.

Thus, the total number of elements of dimension 4 in the product of polytopes under consideration is 613 ($f_4(8) = 613$).

Elements of dimension 3 in the product of a 4 – cross - polytope to a 4 - simplex are the following figures:

1. The product of tetrahedrons of the 4 – cross – polytope onto the vertices of the 4 – simplex, i.e. 80 elements;
2. The product of tetrahedrons of the 4 – simplex onto the vertices of the 4 – cross – polytope, i.e. 40 elements;
3. The product of flat elements of the 4 – cross – polytope onto the edges of the 4 – simplex, i.e. 320 elements;
4. The product of flat elements of the 4 – simplex onto the edges of the 4 – cross – polytope, i.e. 240 elements.

Thus, the total number of elements of dimension 3 in the product of polytopes under consideration is 680 $\left(f_3(8) = 680\right)$.

Elements of dimension 2 in the product of a 4 – cross - polytope to a 4 - simplex are the following figures:

1. The product of flat elements of the 4 – cross – polytope onto the vertices of the 4 – simplex, i.e. 160 elements;

2. The product of flat elements of the 4 – cross – polytope onto the edges of the 4 – simplex, i.e. 320 elements;
3. The product of the edges of the 4 – cross – polytope onto the edges of the 4 – simplex, i.e. 240 elements.

Thus, the total number of elements of dimension 2 in the product of polytopes under consideration is 480 $\left(f_2(8)=480\right)$.

Elements of dimension 1 in the product of a 4 – cross - polytope to a 4 - simplex are the following figures:

1. The product of the edges of the 4 – cross – polytope onto the vertices of the 4 – simplex, i.e. 120 elements;
2. The product of the edges of the 4 – simplex onto the vertices of the 4 – cross – polytope, i.e. 80 elements.

Thus, the total number of the edges in the product of polytopes under consideration is 200 $\left(f_1(8)=200\right)$.

Substituting the obtained values of the number of elements of different dimensions in the product of the 4 – cross - polytope by the 4 - simplex into the Euler - Poincaré equation (2) of Chapter 2, can to obtain

$$40 - 200 + 480 - 680 + 613 - 354 + 122 - 21 = 0.$$

Thus, easy to see that the Euler-Poincaré equation for a given the polytopic prismahedron is satisfied for dimension 8. This proves that the polytope in question exists and has faces in the form of the 4 – cross - polytope.

THE *N* – CROSS – POLYTOPES AS THE MODELS OF THE STRUCTURE OF THE CHEMICAL COMPOUND MOLECULES

Zhizhin (Zhizhin, 2015a) showed the diagram of the tetrahedral molecule B_4Cl_4. Its frame consists of four boron atoms and has the shape of a tetrahedron. Tetrahedral coordination is formed around each boron atom, if an electron pair is placed in the center of each face of frame tetrahedron, and one from the four valence bonds of each boron atom is occupied by a chlorine atom (Gillespie & Hargittai,1975). The dimension this molecule was defined

(Zhizhin, 2015a, b) as dimensions of the polytope formed on set 8 vertices after connected each other boron atoms and each other chlorine atoms, and connected each boron atom with nearest chlorine atom. It is proofed that dimension this polytope equal 4. In this building each vertex there is incidence 4 edges of the polytope. However, this polytope can to attune lead pass on two edges still from each vertex to all remain vertices hold condition uniform of vertices. The result this building (triangulation) shown on Figure 7.

It is evident that this construction is equivalent the 4 – cross - polytope on Figure 9 of Chapter 2. If each vertex on Figure 7 to connected by edges with remained vertices than polytope became of the 7 - simplex (Zhizhin, 2015b).

One will try to explain physically the difference of dimension of the molecule with the same number of atoms (vertices), suggesting the existence of different energy states of the molecule when the number of valence bonds of each atom remains the same. It should be noted that the space between atoms in the molecule is not empty. It is characterized by the presence of

Figure 7. The structure of the molecule B_4Cl_4

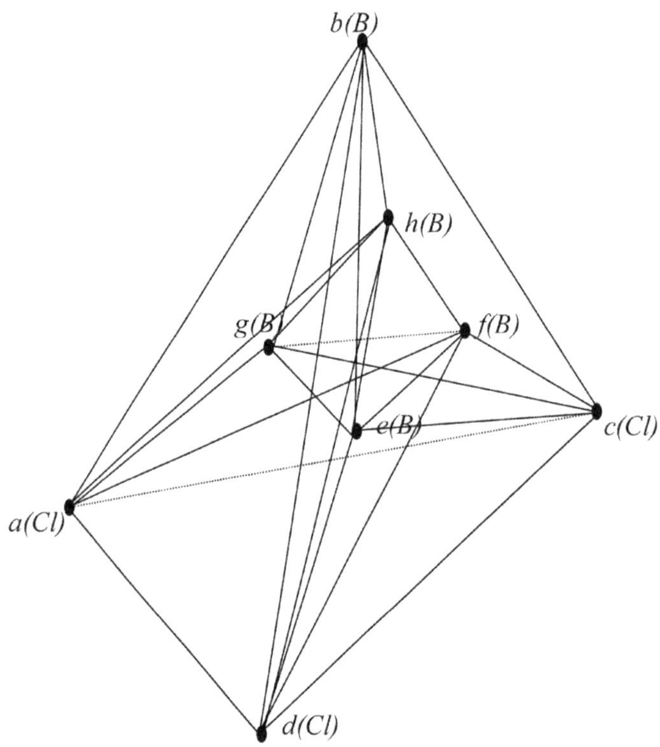

electrons in it, i. e. electron density distribution (Zhizhin, 2015b; Gray, 1965; Sinanoglu, 1965), which can be heterogeneous and has a certain structure. It is obvious that this structure corresponds to the energy state of the molecule and due to the quantum nature of the state this condition change in a discrete manner. When modeling a molecule by polytope, structure of space inside the molecule can be modeled geometrically by a number of edges issuing from each vertex. One can specify only a finite number of options of edges distribution inside the polytope for each given number of vertices under the condition of vertices homogeneity. It appears from this that the number of possible values of the molecule dimension for a given number of vertices is finite and it varies discretely. It follows from the preceding that the more energy is contained in the molecule, the greater the number is the number of edges issuing from each vertex of the corresponding polytope, and more dimensions of the molecule.

The adamantine molecule consisting of 10 atoms of carbon was considered (Zhizhin, 2014; Zhizhin, Khalaj & Diudea, 2016). From each carbon atom in this model as vertex of polytope issue 6 edges. It is proof that dimension this molecule equal 4. Many molecules consist from the ten atoms ($P_4 O_6$, $As_4 O_6$, $Sb_4 O_6$ etc.).

Similar structures have mane molecules with more the number of atoms, for example, $P_4 O_{10}$, $P_4O_6S_4$ and other. Beryllium atoms in crystalline fluoride formed similar construction too. Between atoms of beryllium in this molecule atoms fluorine dispose of line. It does not change the structure of the compound. In all constructions 6 edges issues from each vertex of polytopes. Keeping to the condition of homogeneity of vertices one can increase the number of edges issuing from each vertex to 8 (after triangulation). Then one will get a cross - polytope of dimension 5. Its image is shown on Figure 10 Chapter 2. A further increase of the number of edges issuing from each vertex of the polytope with 10 vertices leads to a simplex of dimension 9, i.e. less by 1 than the number of atoms.

In book of the Gillespie (Gillespie, 1972) shows a dimer structure $NbCl_5$ (Figure 8).

This molecule has 12 atoms (2 Nb atoms and 10 Cl atoms). From Figure 8 shows that in this structure there are 6 pairs of oppositely arranged atoms, between which there is no chemical bond (1 – 6, 3 – 4, 2 – 5, 7 – 8, 10 – 11, 9 – 12) . Bridging atoms edges to form a convex polytope, but these couples, get the 6 – cross - polytope, presented in Figure 11 Chapter 2.

Figure 8. The dimer structure of $NbCl_5$

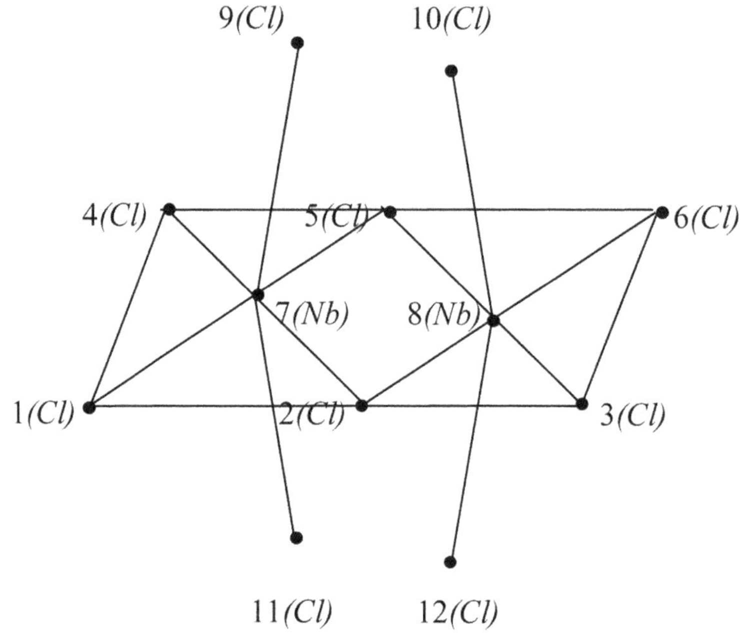

THE POLYTOPE OF GOSSET

One have just considered the question of higher - order polytopes, in which the cross - polytopes can be used as a face. However, for the first time, the possibility of the existence of a polytope with faces from the cross - polytopes in 1900 by T. Gosset (Gosset, 1900) was declared. It was a shrewd note of a man who did not have a mathematical education. It was claimed that one of the semi - regular polytopes of higher dimension could be a polytope of dimension 8, containing 240 vertices, 6720 edges, 60480 two - dimensional faces (triangles), 241920 tetrahedrons, 483840 simplexes 4, 483,840 simplexes of dimension 5, 207360 simplexes of dimension 6, 17280 simplexes of dimension 7 and 2160 of cross - polytopes of dimension 7. There was no evidence of the existence of such a polytope. This work initiated a multitude of works by various authors using Gosset's assertion in their theories, especially after Coxeter referred to this work (Elte, 1912; Coxeter, 1963; Johnson; 1960; Conway & Sloane, 1984; Sadok & Mosseri, 1993; etc.). Coxeter called this polytope Gosset polytope, although Coxeter had no proof of the existence of this polytope. One will not disassemble the essence of these works, including the Coxeter group and the Coxeter - Dynkin diagram, constructed using the Gosset polytope. Using the results on incidence coefficients for faces of polytopes of higher dimension obtained in the previous sections of this book, one will try to clarify the question of the existence of a Gosset polytope. Examples of visualization of the Gosset polytope, available in the literature (see for example Sherk, McMullen & Thompson, 1995), cannot be a proof of its existence. They are rather of purely artistic significance, since they cannot accurately indicate specific geometric figures, list them, determine the nature of their interaction. It can also be added that the classification of polytopes of dimension 8 in the literature (Sherk, McMullen & Thompson, 1995) does not take into account the existence of a whole class of the polytopic prismahedrons of dimension 8, proved in this paper, which is the result of the product of polytopes of different dimensions of each other.

To carry out this analysis, it is convenient to write out the necessary formulas obtained in the previous sections. The number of elements of dimension i in the simplex of dimension n is

$$f_i\left(n\right) = C_{n+1}^{i+1}. \tag{4}$$

The number of facets in the simplex of dimension n is

$$f_{n-1}(n) = C_{n+1}^{n} = n + 1. \tag{5}$$

The number of elements of dimension i in the cross - polytope of dimension n is

$$f_i\left(n\right) = 2^{1+i} C_n^{n-i-1}. \tag{6}$$

The number of facets in the cross - polytope of dimension n is

$$f_{n-1}(n) = 2^n. \tag{7}$$

The incidence coefficient of an element of dimension i with respect to an element of dimension $i + 1$ in a simplex of dimension n is

$$k^s_{b(i)}(n) = n - i \tag{8}$$

The incidence coefficient of an element of dimension i with respect to an element of dimension $i + 1$ in a cross – polytope of dimension n is

$$k^{c}_{b(i)}(n) = 2(n - i - 1). \tag{9}$$

So, Gosset asserts that 2160 the cross - polytopes with dimension 7 are included in his polytope, but in each the 7 - cross-polytope, according to (7), there are 128 the simplexes of dimension 6. Consequently, set the cross – polytopes with dimension 7 is included 2160 · 128 = 276480 the simplexes with dimension 6. Gosset asserts that 17280 the simplexes with dimension 7 are included in his polytope, but in each 7 – simplex, according to (5), there are 8 simplexes with dimension 6. Consequently, the simplexes with dimension 7 is included 17280 · 8=138240 the simplexes with dimension 6. However, each the 6 – simplex is included in two polytopes of dimension 7 in area of the cross – polytopes (see (9)) and in area of simplexes (see (8) ($k^{s}_{b(6)}(8) = 2, k^{c}_{b(6)}(8) = 2$). Consequently, general number of the 6 – simplexes in area of the 7- cross – polytopes equal 276480/2 = 138240 and in area of the 7 – simplexes general number of the 6 – simplexes equal 138240/2 = 69120. Thus, general number of the 6 – simplexes in the polytope of Gosset equal 207360. This coincidence with data of Gosset.

In each 6 – simplex in area the 7 – cross – polytopes is included 7 simplexes with dimension 5 (see (5), $f_5(6) = 7$). Consequently, in 138240 simplexes with dimension 6 in area the 7 – cross – polytopes are included 138240 · 7 = 967680 simplexes with dimension 5. However, in this area each the 5 - simplex appertain four the 6 – simplexes ($k^{c}_{b(5)}(8) = 4$, see (9)). Consequently, general number of the 5 – simplexes in area of the cross – polytopes equal 967680/4 = 241920. In each 6 – simplex in area of the 7 – simplexes are included 7 simplexes with dimension 5 ($f_5(6) = 7$, (5)). Consequently, in 161280 simplexes with dimension 6 in area of the 7 – simplexes are included 161280 · 6 simplexes with dimension 4. However, in this area each 4 – simplex appertain four the 5 – simplexes ($k^{s}_{b(4)}(8) = 4$, see (8)). Consequently, general number of the 4 – simplexes in this area equal 161280 · 6/3 = 161280. Thus, general number of the 5 – simplexes in polytope of Gosset equal 403200. This number not coincidence with data of Gosset, in which general of the 5 – simplexes equal 483840.

In each 5 – simplex in area the 7 – cross – polytopes are included 6 simplexes with dimension 4 (see (5), $f_4(5) = 6$). Consequently, in 241920 simplexes with dimension 6 in area the 7 – cross – polytopes are included 241920 · 6 = 1451520 simplexes with dimension 4. However, in this area each 4 - simplex appertain six the 5 – simplexes ($k^c_{b(4)}(8) = 6$, see (9)). Consequently, general number of the 4 – simplexes in area of the cross – polytopes equal 1451520/6 = 241920. In each 5 – simplex in area of the 7 – simplexes are included 6 simplexes with dimension 4 ($f_4(5) = 6$, (7)). Consequently, in 161280 simplexes with dimension 5 in area of the 7 – simplexes are included 161280 · 6 = 967680 simplexes with dimension 4. However, in this area each 4 – simplex appertain four the 5 – simplexes ($k^s_{b(4)}(8) = 4$, see (8)). Consequently, general number of the 4 – simplexes in this area equal 967680/4 = 241920. Thus, general number of the 4 – simplexes in polytope of Gosset equal 483840. This number coincidence with data of Gosset, in which general number of the 4 – simplexes equal 483840.

In each 4 – simplex are included 5 tetrahedrons. Consequently, in area the cross – polytopes in set the 4 – simplexes are included 5 · 241920 tetrahedrons. However, in this area (cross – polytopes) each tetrahedron appertain 8 simplexes with dimension 4 ($k^c_{b(3)}(8) = 8$). Consequently, general number of tetrahedrons in area of the cross – polytopes equal 5·241920/8= 151200. In area the simplexes in set of the 4 – simplexes are included 5·241920 tetrahedrons. However, in this area (simplexes) each tetrahedron appertain 5 the 4 – simplexes ($k^s_{b(3)}(8) = 5$). Consequently, general number tetrahedrons in area of the simplexes equal 241920. Thus, general number tetrahedrons in polytope of Gosset equal 393120. This number not coincidence with data of Gosset, in which general number of tetrahedron equal 241920.

In each tetrahedron are included 4 flat triangles. Consequently, in 151200 tetrahedrons in area of the cross – polytopes are included 151200·4 flat triangles. However, in area of the cross – polytopes each flat triangle appertain 10 tetrahedrons ($k^c_{b(2)}(8) = 10$). Thus, general number of flat triangles in area of the cross – polytopes equal 151200·4/10 = 60480. In area of the simplexes in tetrahedron are included 241920·4 flat triangles. However, in area of the simplexes each flat triangle appertain 6 tetrahedrons ($k^s_{b(2)}(8) = 6$). Consequently, general number flat triangles in area of the simplexes equal 241920·4/6 = 161280. Thus, general number of flat triangles in polytope of Gosset equal 221760. This number not coincidence with data of Gosset, in which general number of flat triangles equal 60480.

In each flat triangle are included 3 edges. Consequently, in set of flat triangles in area of the cross – polytopes are included 181440 edges. However, in area of the cross – polytopes each edges appertain 12 flat triangles ($k^{c}_{b(1)}(8) = 12$). Thus, in area of the cross – polytopes are 181440/12 = 15120 edges. In area of the simplexes are $3 \cdot 161280 = 483840$ edges. However, in area of the cross – polytopes each edges appertain 7 flat triangles ($k^{s}_{b(1)}(8) = 7$). Thus, in area of the simplexes are 483840/7 = 69120 edges. Consequently, in polytopes of Gosset are 84240 edges. This number not coincidence with data of Gosset, in which the number of edges equal 6720.

In each edge are included two vertices. Consequently, in area of the cross – polytopes are $15120 \cdot 2 = 30240$ vertices. However, each vertex in area of the cross – polytopes appertain to 14 edges. Consequently, general number of vertices in area of the cross – polytopes equal 30240/14 = 2160. Each vertex in area of the simplexes appertain to 8 edges. Thus, general number of vertices in area of the simplexes equal 69120/4 = 17280. Consequently, general number of vertices in polytope of Gosset equal 19440. This number not coincidence with data of Gosset, in which general number of vertices equal 240.

Thus, the statement about the existence of a polytope of dimension 8 with 240 vertices and facets of 2160 cross - polytopes and 17280 simplexes, put forward by Gossett in 1900, should be considered erroneous. Assuming the existence of a polytope of dimension 8 with the indicated quantities of facets from the cross - polytopes and the simplexes, using the geometric properties of polytopes of higher dimension, investigated in this book, one have arrived at a contradiction. The calculated values of the number of polytopes of dimension less than the dimension of the facets turned out to be different from the values indicated by Gosset. In particular, the number of vertices turned out to be much larger than 240. Moreover, the obtained values of the quantities of faces of different dimensions on satisfy the Euler - Poincaré equation, since the substitution of these values in this equation (equation (2) Chapter 2) does not lead to its solution for dimension 8. This means that the obtained as a result of the calculation of the polytope does not exist, as there is no polytope with the values of faces of different dimensions given by Gosset. In this connection, the use of the Gosset polytope for describing structures in generalized crystallography (Sadoc & Mosseri, 1993), ball packings (Coxeter, Few & Rodgers,1959; Coway & Sloane, 1984) and coding theory (Lowenstein, 1983) requires revision. The main reason for the erroneousness of Gosset's assertion is that he did not take into account a substantial increase in incidence of faces with increasing polytope dimension. This geometric property of polytopes of higher dimension was investigated in this book and was previously unknown.

CONCLUSION

The structure of the n – cross - polytope is studied in detail for large values of n. Confirmed by the construction that the n – cross - polytope is a simplicial polytope, all its faces of dimension less than n are the simplexes. Although among the set of tetrahedrons entering the boundary complex of the n – cross - polytope it is possible to distinguish the cross - polytopes of dimension less than n, but the possible numbers of these faces obtained do not satisfy the Euler - Poincare equation.

A general analytic expression is expressed for the incidence coefficient of faces of any dimension n - cross - polytope with respect to its facets. It is essential that, according to this expression, the value of this coefficient increase sharply with increasing n and the dimension of the face. The n – cross - polytope can be the face of the polytope, if this polytope is a polytopic prismahedron. An example of such a polytope can be the product of the 4 – cross - polytope by the 4 - simplex. An image of this polytopic prismahedron is constructed, all the faces included in its boundary complex are listed. Examples of specific chemical compounds whose molecules are described by the n – cross - polytopes with values of n equal to 4, 5, and 6 are given. Based on the analytical expressions obtained in this book for the numbers of faces of various dimensions in the n – cross - polytope and the n - simplex, and the incidence coefficients of the faces in these polytopes, the question of the possibility of the existence of a Gosset polytope was considered. As the initial condition, the numbers of faces of dimension 7 indicated by Gosset for the polytope of dimension 8 are taken. Then, taking into account the incident coefficients of the faces (according to the law obtained in this book), the numbers of faces of lower dimension, up to vertices, are determined successively. It turned out that the obtained numbers as a whole differ significantly from the numbers of faces of different dimensions indicated by Gosset. In this case, the numbers of faces obtained in this way do not satisfy the Euler-Poincare equation. This proves that Gosset's polytope does not exist. In connection with this result, it is required to revise many works in the theory of crystallography and coding theory, based on the hypothesis of the existence of a Gosset polytope.

REFERENCES

Conway, J. H., & Sloane, N. J. A. (1984). *The Leech Lattice, Sphere Packings, and Related Topics*. New York: Springer - Verlag.

Coxeter, H.S.M. (1963). *Regular Polytopes*. London: John Wiley & Sons, Inc.

Coxeter, H. S. M., Few, L., & Rodgers, C. A. (1959). Covering space with equal spheres. *Mathematics*, *6*, 147–157.

Elte, E. L. (1912). *The semi-regular polytopes of the hyperspace*. Groningen: University of Groningen.

Gillespie, R. J. (1972). *Molecular Geometry. London, New York, Cincinnaty*. Toronto: Van Nostrand Reinhold Company.

Gillespie, R. J., & Hargittai, I. (1975). *The VSEPR Model of Molecular Geometry*. London: Allyn and Bacon.

Gosset, T. (1900). On the regular and semi - regular figures in space of *n* – dimensions. *The Messenger of Mathematics*, *29*, 43–48.

Gray, H. B. (1965). *Electrons and Chemistry Bounding*. Amsterdam: W.A. Benjamin Ink.

Grunbaum, B. (1967). *Convex Polytopes*. London: Springer.

Johnson, N. W. (1960). *The Theory of Uniform Polytopes and Honeycombs* [Ph.D. Dissertation]. University of Toronto.

Lowenstein, V. I. (1983). *Bounds for packings of metric spaces and some of their applications, Problems of Cybernetics*. Moscow: Science.

Poincare, A. (1895). Analysis situs. *J. de é Ecole Polytechnique, 1,* 1 – 121.

Sadoc, J. F., & Mosseri, R. (1993). The E_8 lattice and quasicrystals: Geometry, number theory and quasicrystals. *Journal of Physics. A, Mathematical and General*, *26*(8), 1789–1809. doi:10.1088/0305-4470/26/8/009

Sherk, F. A., McMullen, P., & Thompson, A. C. (1995). *Kaleidoscopes: Selected Writings of H.S.M. Coxeter*. Asia Ivic Weiss, Wiley-Interscience Publication.

Sinanoglu, O. (1965). *Modern Quantum Chemistry*. London: Academic Press.

Sloane, N. J. A. (1984). Sphere Packings. *Scientific American*, *3*, 72–82.

Stringham, W. I. (1880). Regular figures in n- dimensional space. *American Journal of Mathematics*, *3*(1), 1–14. doi:10.2307/2369441

Zhizhin, G. V. (2014a). *World – 4D*. St. – Petersburg: Polytechnic Service.

Zhizhin, G.V. (2014b). On high dimensionalities in nature, *Biosphere, 6*(4).

Zhizhin, G. V. (2015a). The dimensionality of supramolecular compounds. *Biosphere*, *7*(2), 149–154.

Zhizhin, G.V. (2015b). Dimensions of compounds in supramolecular chemistry. *Int. J. Chem. Modeling, 7*(2).

Zhizhin, G.V. (2016). N – prismahedrons and their dual polytopes. *International Journal Chemical Modeling, 5*(4).

Zhizhin, G. V., & Diudea, M. V. (2016a). Space of nanostructures. In M. Putz (Ed.), *Sustainable nanosystems, development, properties, and application*. Hershey PA: IGI Global.

Zhizhin, G. V., Khalaj, Z., & Diudea, M. V. (2016). Geometrical and topological dimensions of the diamond. In A. R. Ashafi & M. V. Diudea (Eds.), *Distance, symmetry and topology in carbon nano-materials*. London, New York: Springer. doi:10.1007/978-3-319-31584-3_12

Ziegler, G. (1995). *Lectures on polytopes*. New York: Springer. doi:10.1007/978-1-4613-8431-1

KEY TERMS AND DIFINITIONS

Incidence in Polytopes: Incidence in polytopes define the number of elements of higher dimension the given element of lower dimension belongs.

Polytope With Factor – Structure: Polytope in which vertices are polytopes with same dimension k, and edges are polytopes with dimension $k - 1$.

Simplicial Polytope: Polytope with dimension n in which all faces with dimension less n are simplexes.

Chapter 7
Scaling in the Process of Hierarchical Filling of *n* – Dimensional Space

ABSTRACT

The hierarchical filling of the n - dimensional space with geometric figures is studied, accompanied by a process of discrete similar changes in their dimensions, i.e. process of scaling. The scaling process in these fillings does not depend on time and is determined only by the geometric characteristics of the figures, which are preserved when their size is changed. Two possible ways of hierarchical filling of space are defined, under which the original figure incrementally increases its size fills the space. Investigations of the hierarchical filling of concrete geometric figures of a plane, three -dimensional space, four - and five - dimensional spaces are carried out. The denominator of geometric progressions characterizing sequences of figures in the process of scaling are determined depending on the shape of the figure and its dimension.

INTRODUCTION

For the first time the question of the hierarchical filling of the plane by geometric figures was posed by Kepler (Kepler, 1611). The hierarchical arrangement of any triangles on the plane was considered of Zhizhin (Zhizhin, 1993) and hierarchical arrangement special rhombs on the plane was considered of Penrose (Penrose, 1974). The hierarchical arrangement of regular pentagons

DOI: 10.4018/978-1-5225-6968-8.ch007

on the plane was considered of Mackay (Mackay, 1981, 2006). In these works, the hierarchy of a structure was understood in the sense that at each stage of its construction, an exact replica of some initial shape (not just the pentagon) is used from the subnets. At the same time, the structures from sub - sources are not dense, i.e. between the subordinates there is an empty space of some form. However, more accurately the notion of hierarchy corresponds to structures in which there are several levels corresponding to the subtle forms of a certain figure that differ in size (Zhizhin, 2010, 2012, 2014). In these hierarchical structures, the question of scaling naturally arises, i.e. scale change in the shape of some original figure. This concept of scaling differs from the process of scaling (or scale invariance) in physics, since in this concept there is no temporal coordinate. The scaling process in physics consists in preserving the equations of the process when the distances and time intervals are changed by the same number of times (Prokhorov, 1992). The scaling process in this case is only a variant of the long - known self -similarity of the process (Barenblatt, 1982). One will consider hierarchical structures that have different levels of organization, differing in the sizes of such figures. Before considering the hierarchical filling of n - dimensional spaces, it is reasonable to turn to the hierarchical filling of a two - dimensional plane (Zhizhin, 2010).

HIERARCHICAL FILLING OF TWO - DIMENSIONAL FLAT

Let the figure that one fill the plane be a regular pentagon $A_0B_0C_0D_0E_0$ (Figure 1).

Continuing its sides before crossing into the outer region, can to get a pentagon similar to the original one $A_1B_1C_1D_1E_1$. Having done the same with a pentagon $A_1B_1C_1D_1E_1$, can to get an even larger regular pentagon $A_2B_2C_2D_2E_2$, and so on. By joining the adjacent vertices of a polygon $A_0B_0C_0D_0E_0$, skipping one next vertex, can to obtain a polygon $A_{-1}B_{-1}C_{-1}D_{-1}E_{-1}$ inside the original and then an even smaller polygon $A_{-2}B_{-2}C_{-2}D_{-2}E_{-2}$, and so on. Considering such triangles formed from the intersection of lines, can to conclude that $\Delta D_0D_{-1}C_0 \sim \Delta E_1D_1C_1$, and hence

Figure 1. The regular pentagon

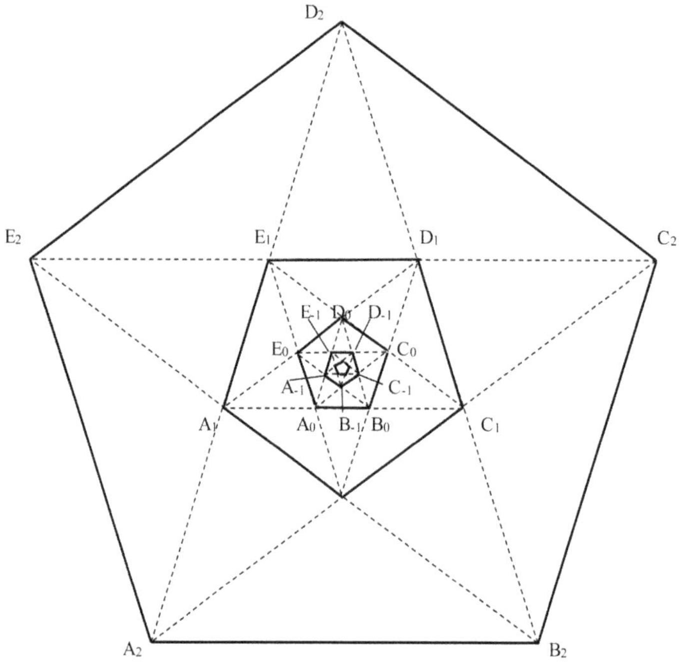

$$\frac{D_0C_0}{2E_1D_0 + D_0C_0} = \frac{D_{-1}C_0}{D_1C_1}, \tag{1}$$

but

$$\frac{D_{-1}C_0}{D_1C_1} = \frac{D_{-1}C_0}{D_1C_0 + B_0C_0} = \frac{D_{-1}C_0}{D_0C_0 + E_1D_0} = \frac{D_{-1}C_0 / E_1D_0}{1 + D_0C_0 / E_1D_0} = \frac{D_0C_0 / E_1D_0}{\left(1 + D_0C_0 / E_1D_0\right)^2}. \tag{2}$$

One denote the relation $E_1D_0 / D_0C_0 = \tau$, then, substituting (2) in (1), after the transformations there is

$$\tau^2 - \tau - 1 = 0. \tag{3}$$

The value has only the positive root of equation (3)

$$\tau = \frac{1+\sqrt{5}}{2} = 1,618033989 \quad (4)$$

This root characterizes the golden section of the segment E_1C_0 into two parts E_1D_0, D_0C_0. It is obvious that such divisions of segments exist in every regular pentagon that arises on the plane as a result of construction. Since the size of a regular polygon it is characterized by the length of its side, the sequence of pentagons can be compared with a series of lengths of the sides of the pentagons. The ratio of the sides of regular pentagons that are close in size is constant and equal

$$\frac{E_1D_1}{D_0C_0} = \frac{E_1D_0 + D_0C_0}{D_0C_0} = \tau + 1. \quad (5)$$

If one consider the length of the side of the original pentagon as a unit, then the sequence of lengths of the sides of the increasing pentagons corresponds to the series

$$\ldots(\tau+1)^{-3}, (\tau+1)^{-2}, (\tau+1)^{-1}, 1, \tau+1, (\tau+1)^{2}, (\tau+1)^{3}, \ldots \quad (6)$$

where the terms of the series after 1 correspond to pentagons spaced outside the original pentagon, and the terms of the series to one correspond to the pentagons located inside the original pentagon. The sequence (6) is an increasing geometric progression with the denominator $\tau + 1$.

Such a construction models, for example, the growth of petals of flowers, which, as a rule, have a smaller size at the center and increasing in size away from the center, or the growth of cabbage leaves that are small size in center and large in size at the periphery. Thus, with the help of a regular pentagon it is possible to construct, so - called (Zhizhin, 2010), an increasing geometric variety modeling the growth of living systems. Between the pentagons in this manifold, a discrete scaling is performed and the similarity coefficient $\tau + 1$ can be considered a scaling function. A regular pentagon is not the only polygon possessing this property. Figure 2 shows the growing geometric variety with an example of a regular heptagon (a flower - a seven - color).

Figure 2. The increasing geometric variety on bases of regular heptagon

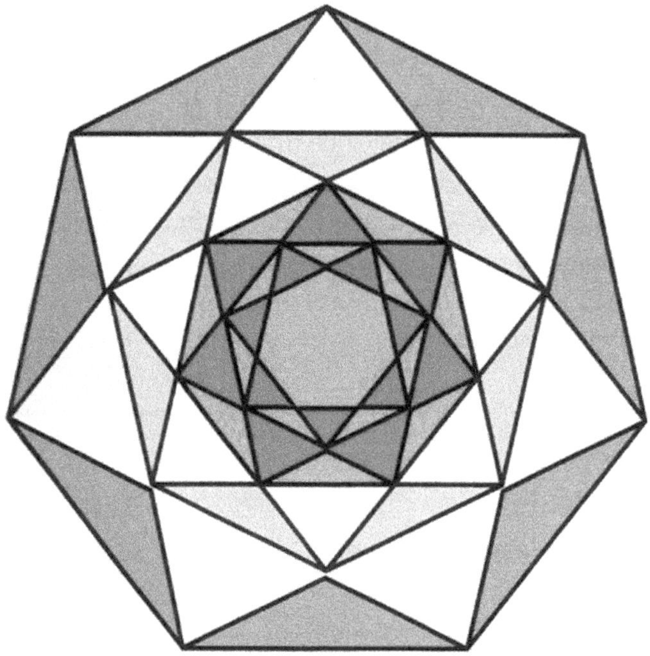

Theorem 1

Each regular n - gon for $n \geq 5$ can fill the plane with a growing geometric manifold with division of distances between adjacent vertices in the polygon with respect to $d_n = 0,5\left(2\cos^2\frac{\pi}{n} - 1\right)^{-1}$, and the sides of the polygons thus form an increasing geometric progression with the denominator $q_n = \dfrac{\cos\left(\pi / n\right)}{\cos\left(2\pi / n\right)}$.

Proof: Assume that the original regular n - gon ($n \geq 5$) has vertices $B_0, C_0, D_0, E_0, \ldots$ (Figure 3).

Through the vertices B_0, C_0, D_0, E_0 there passes a circle with a radius R and the center O into which this n - gon is inscribed. Continuing the sides C_0B_0, D_0E_0 to the intersection, can to get the vertex D_1 of the polygon of the

Figure 3. Filling the plane with the correct polygon in the "growing" mode

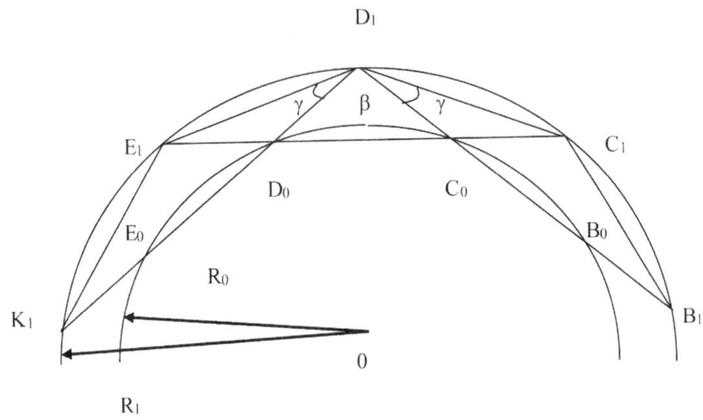

nearest larger size. Through this vertices there passes a circle with a radius R_1 and center O. Continuing the sides C_0B_0, D_0E_0, D_0C_0 to the intersection with circle with radius R_1 can to get also vertices C_1, B_1, E_1, K_1 of n – gon inscribed in a circle of radius R with center O. The angle $K_1D_1B_1$, as an inscribed angle, resting on an arc K_1B_1, is equal to half the central angle, resting on the same arc,

$$\angle K_1D_1B_1 = \beta = \frac{1}{2}\left(2\pi - 4\frac{2\pi}{n}\right) = \pi\frac{n-4}{n}.$$

Thus,

$$\angle E_1D_1D_0 = \angle C_1D_1C_0 = \angle D_1E_1D_0 = \angle D_1C_1C_0 = \frac{\alpha-\beta}{2} = \frac{\pi}{n} = \gamma. \qquad (7)$$

Triangles $E_1D_1D_0, D_1C_1C_0$ are similar of triangle $E_1D_1C_1$. Therefore,

$$\frac{E_1D_1}{D_0C_0 + 2E_1D_0} = \frac{E_1D_0}{E_1D_1}$$

that is, by introducing $d_n = \dfrac{E_1D_0}{D_0C_0}$, can to obtain

$$\left(\frac{E_1D_1}{D_0C_0}\right)^2 = d_n\left(2d_n + 1\right). \tag{8}$$

But, taking into account (7) $E_1D_1 = E_1D_0 2\cos\frac{\pi}{n}$ or

$$\frac{E_1D_1}{D_0C_0} = 2d_n \cos\frac{\pi}{n}. \tag{9}$$

Substituting (9) into (8), after transformations there is

$$d_n = \frac{1}{2}\left(2\cos^2\frac{\pi}{n} - 1\right)^{-1}. \tag{10}$$

The first assertion of Theorem 1 is proved.

The second assertion of Theorem 1 follows from (8), (10). It is true, from (8) follows that

$$\frac{E_1D_1}{D_0C_0} = \sqrt{d_n\left(2d_n + 1\right)}.$$

Substituting in this equation (10), can to find

$$\frac{E_1D_1}{D_0C_0} = \frac{R_1}{R_0} = q_n = \frac{\cos\left(\pi / n\right)}{\cos\left(2\pi / n\right)}. \tag{11}$$

Consequently, the sides of the n - gons form an increasing geometric progression with the denominator $q_n = \frac{\cos\left(\pi / n\right)}{\cos\left(2\pi / n\right)}$. Theorem 1 is proved.

It is of interest to compare the arbitrary term of the geometric progression defined by the theorem 1 with the two preceding members of this progression. Denote by $S_{n,k}$ the length of the side of the regular n - gon at the k - th step of the increase from the initial n - gon. By the geometric progression, one have $S_{n,k} = q_n S_{n,k-1}$. Imagine

$$S_{n,k} = S_{n,k-1} + (q_n - 1)S_{n,k-1} = S_{n,k-1} + q_n(q_n - 1)S_{n,k-2}. \qquad (12)$$

It follows from (12) that the terms of the geometric progression $S_{n,k}$ represent the Fibonacci series (i.e. each term of the series is the sum of the two preceding ones) only then, when $q_n(q_n - 1) = 1$, i.e. then when

$$q_n = \frac{1+\sqrt{5}}{2} = \tau = 1,681033989.$$

But can there be such a value of q_n? Table 1 shows the results of calculating d_n and q_n as a function of n from the obtained formulas (10), (11).

It follows from the table that the sequence of lengths of sides of regular *n* - gons for any *n* is not a Fibonacci series. In addition, it follows from the table that as *n* increases, the denominator of the geometric progression decreases, tending to unity.

This means that the second term in (12) decreases rather rapidly, asymptotically tending to zero. From Figure 1, the growth of regular pentagons implies that the Fibonacci series consists of the sides of regular pentagons and, alternating with them, the sides of star pentagons (for example, $E_0E_{-1} + E_{-1}D_{-1} = E_0D_0$, etc.). However, it remains unclear the meaning of

Table 1. The functions d_n and q_n at number sides of polygon

n	d_n	q_n	$q_n(q_n - 1)$	q_n^*
3	-	-	-	2
4	-	-	-	1,41421356
5	1,6180033989	2,618033989	4,236067	1,2360679
6	1	1,73205	1,2679472	1,15470538
7	0.801937735	1,445041867	0,6431041	1,10991626
8	0,707106781	1,306562965	0,4005438	1,0823922
9	0,652703647	1,2266816	0,27806614	1,06417772
10	0,618033989	1,1755705	0,2063955	1,05146222
11	0,594351144	1,140551777	0,16030658	1,04221711
12	0,577350269	1,115355071	0,12866186	1,03527618
13	0,56468078	1,096544366	0,10586518	1,02992783

the summation of clearly inconvenient figures, which are convex and star - shaped pentagons.

In all other cases, for any *n* sequences of sides of convex polygons, sequences of sides of star polygons, sequences of alternating sides of convex and star polygons do not lead to the widely known Fibonacci series (Vorobiev, 1984). Instead of the Fibonacci series, for a sequence of sides of convex regular polygons of interest, a more complex series is realized, characterized by a decrease in the influence (memory) of the selected member of the series, lagging behind it by two steps of the other member of the series.

For a regular triangle and a square, can to suggest a slightly different way of obtaining a growing geometric variety. Consider, for example, the right triangle. From the center of the triangle draw the bisectors of the angles of the triangle and continue them outside the triangle (see Figure 4). To each of the bisectors one draw perpendiculars passing through the vertices of the triangle.

To each of the bisectors can to draw perpendiculars passing through the vertices of the triangle. Three perpendiculars, intersection, form the right triangle of the new generation. Having carried out the same procedures with a new triangle, can to get the right triangle of the next generation. Obviously, such constructions can be carried out with a regular polygon for any number of sides. Let us study this method of obtaining a growing geometric variety

Figure 4. The increasing geometric variety on bases of regular triangle

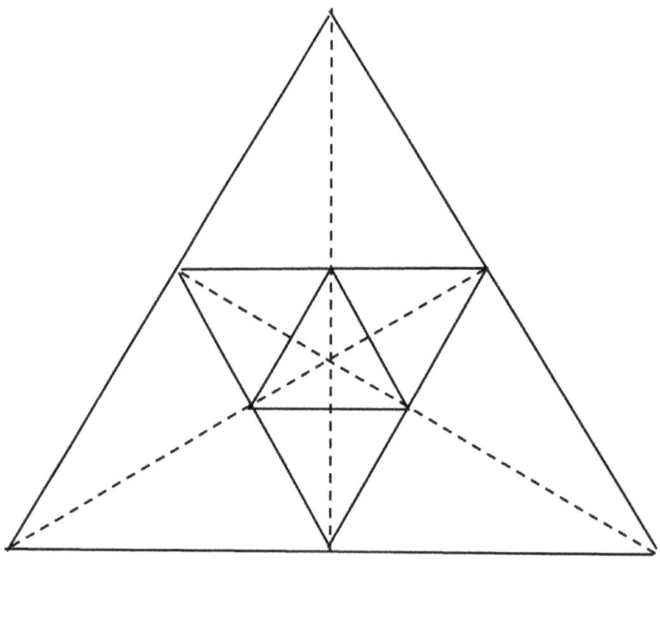

Figure 5. Filling the plane with the correct polygon in the "growing" mode of second way

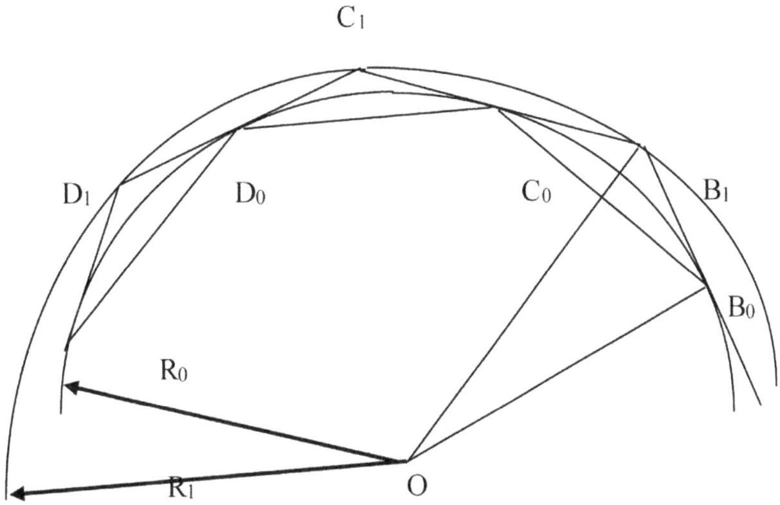

in the general case. Assume again that the original regular n - gon ($n \geq 3$) has the vertices $B_0, C_0, D_0, E_0, \ldots$ (see Figure 5).

Through the vertices B_0, C_0, D_0, E_0 there passes a circle with radius R_0 and center O, into which this n - gon is inscribed. One draw the bisectors of the angles of the n - gon, which naturally pass through the center of the circle. To these bisectors one draw perpendiculars passing through the vertices of the polygon. Crossing, the perpendiculars form a n - gon of a new generation with the vertices $B_1, C_1, D_1, \ldots$. They lie on a circle with radius R_1 and center O. One consider two triangles $\Delta\ OB_1B_0$ and ΔOFB_0 . These two triangles are rectangular with a common angle $\angle B_0OB_1 = \pi / n$. From the similarity of these triangles it follows that

$$\frac{B_1B_0}{FB_0} = \frac{C_1B_1}{C_0B_0} = \frac{1}{\cos\left(\pi / n\right)} = q_n^* = \frac{R_1}{R_0}. \tag{13}$$

In the last column of Table 1, the values of the geometric progression denominator q_n^* are shown for this variant of constructing a growing geometric manifold as a function of the number of sides of a regular polygon n. From the consideration of Table 1 it follows that the denominator of the geometric progression q_n^* increases n very rapidly with increasing 1. This means that, with $n > 5$ the increasing geometric variety practically ceases to expand; remains in the finite domain. This property is also observed in the first method of constructing a growing geometric manifold, but the restriction of its growth in this case occurs for sufficiently large n ($n > 12$). Therefore, it is better to use the second method for constructing a growing geometric manifold for small values. It can be seen that when constructing a growing geometric variety in the second way, the lengths of the sides of polygons of different generations also do not form a Fibonacci sequence.

It should be emphasized that the introduced concept of a growing geometric variety, in spite of its relative simplicity, has a number of properties that allow us to consider a growing geometric variety as a model of growing living systems. Any regular n - gon generates a growing geometric variety. In this case, at each growth stage a new regular n - gon of a larger size is formed and triangles forming a new n - gon, similar to the triangles making n - gon in the previous stage of growth, also appear. A growing geometric variety generates an infinite discrete series of proportions, in which the golden section is only a particular case. It is interesting that, in spite of the fundamental possibility of filling the entire plane with the growing geometric variety with an unlimited increase in the number of stages of growth, the rate of increase in the size of figures decreases with increasing number, and the growing geometric variety remains in the finite domain. The center of a growing geometric variety is a singular point, since when moving towards the center, there is an unlimited reduction of figures that cannot continuously turn into a point. The concept of singular points in tiling of a plane it was introduced by Grunbaum and Shephard (Grunbaum & Shephard, 1987).

Each proportion in an infinite number of proportions corresponds to some geometric progression, and, and this is essential, the denominator of the progression depends on the number n. The algorithm for filling the plane with an infinite number of similar figures (regular n -gons) increasing their size for moving away from some point of the plane (center of growth) is constructed. This algorithm is consistent with the views of the ancient Chinese sages about the living world. At the views an organism possessing the principle of scale self - similarity, in which each individual the thing consists of parts similar to it as a whole, and at the same time this thing itself is one of the parts of the whole world and is structurally similar to it (Vladimirov, 2002).

A comparison of the resulting growing geometric manifolds with the forms of living systems such as flowers shows the similarity of the regularities of growing geometric varieties and colors having several layers of petals and a limited size. Especially close are the characteristics of growing geometric varieties to the colors of the layers of petals, which are located almost in the same plane. For example, Eupomatia bennettii flowers have about 8 petals in each layer (see Takhtadjyan, 1980). Moreover, the ratio of the radius of the second layer to the first (later) is 1.36, which is quite close to the value $q_n = 1,306$ given in the Table 1 for $n = 8$. Similar close values of the ratio of radii to the corresponding values are found in dahlias and chrysanthemums. In all cases (see Takhtadjyan, 1980; Novak, 1976) in flowers of plants with a large number of petals, when passing from one layer of petals to another, the earlier, the diameter of the layer of petals increases insignificantly, and the less, the more the number of petals in the layer. This completely agrees with the dependence of the basis of the geometric progression q_n on n, presented in Table 1. This is completely inconsistent with the assumption of the subordination of the laws of plant growth to the Fibonacci series and the associated golden section, since in this case the ratio of the diameters of the petal layers should be a significant value of 2.618033989 and should not depend on the number of petals in the layer.

HIERARCHICAL FILLING OF THREE - DIMENSIONAL SPACE

Hierarchical Filling of Space With a Tetrahedron

In space, the faces of the tetrahedron do not intersect when they continue to form a larger tetrahedron. Therefore, in order to construct a similar tetrahedron, it is necessary to extend the first method of constructing such a polygon to spaces. The bisectors of angles in the plane case are transformed in space into planes that cut the faces along the bisectors of the angles passing through the two vertices of the tetrahedron. Three such planes pass through each vertex of the tetrahedron. They form four perpendiculars, coming from each vertex perpendicular to the opposite face.

Draw to each of these four perpendiculars a perpendicular plane passing through each vertex, respectively. Crossing, these planes form a new tetrahedron of a larger size. This is a consequence of the duality property of the tetrahedron to itself. Continuing this process, we obtain a sequence of increasing tetrahedron with a single center.

Can to denote the vertices of the original tetrahedron O_1, O_2, O_3, O_4 and the vertices of the larger tetrahedron *A, B, C, D* in this construction (see Figure 6).

One denote the midpoints of the edges of the outer tetrahedron *BD*, *CB* and *CD*, respectively, *S*, *T* and *E*. Let the length of the edge of the outer tetrahedron be "*a*".

One consider the triangle *AEB*. The length of the side *AB* of this triangle is, by assumption, "*a*". The lengths of the sides *AE* and *EB* are equal to $a\frac{\sqrt{3}}{2}$ as heights in the right triangle. The O_4O_3 segment in the triangle *AEB* is parallel to the side *AB*. The length of the segments $O_3B = AO_4$ on the sides *AE, EB* in the triangle *AEB* is found from the triangle *CBD*.

Considering triangles TBO_3 and *CBE* in triangle *DCB*, can to find $\frac{BE}{TB} = \frac{CB}{BO_3}$.

Since $CB = a$, $TB = a/2$, $BE = a\frac{\sqrt{3}}{2}$, from the last equality can to find

Figure 6. Hierarchical filling of space with a tetrahedron

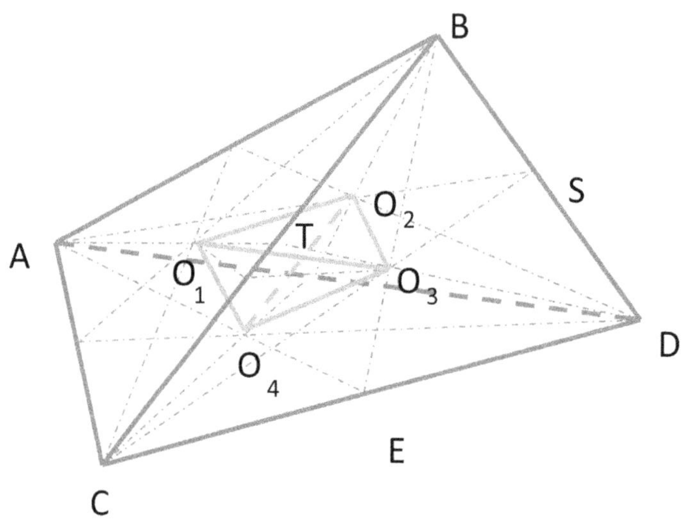

$$BO_3 = AO_4 = \frac{a}{\sqrt{3}}.$$

Now, considering similar triangles EO_4O_3, EAB, there is $\frac{O_4O_3}{AB} = \frac{EO_3}{EB}$.

Since $AB = a, EO_3 = a\frac{\sqrt{3}}{2} - \frac{a}{\sqrt{3}}, EB = a\frac{\sqrt{3}}{2}$ from the last equality can to obtain

$$O_4O_3 = \frac{a}{3}.$$

Thus, the lengths of the edges of tetrahedrons that hierarchically fill the space form a geometric progression with the denominator of the progression $q_T = 3$.

Hierarchical Filling of Space With a Cube and an Octahedron

The edges of the cube, when extended, do not give additional edges, which could lead to the formation of a larger cube. Therefore, to construct a polyhedron external to the cube, on the basis of the characteristics of the cube alone, the first method of construction is applicable.

We draw through each vertex of the cube perpendicular to the diagonal of the cube passing through this vertex, the plane. Crossing, these planes form a polyhedron. Since the vertices of the cube are 8, the resulting polyhedron will be an octahedron that has exactly 6 vertices, i.e. as much as the faces of the cube. This is a manifestation of the duality property of a cube and an octahedron. Therefore, if a polyhedron is constructed outside the octahedron by the same principles, then a cube is obtained. Thus, in this case the sequence of polyhedron consists in alternating cubes and octahedron: a cube gives rise to an octahedron, an one gives rise to a cube. In order to determine the changes in the lengths of the edges of polyhedron in this sequence, can to turn to Figure 7, which depicts an octahedron whose focal points should be the vertices of an inscribed cube.

Figure 7.The octahedron

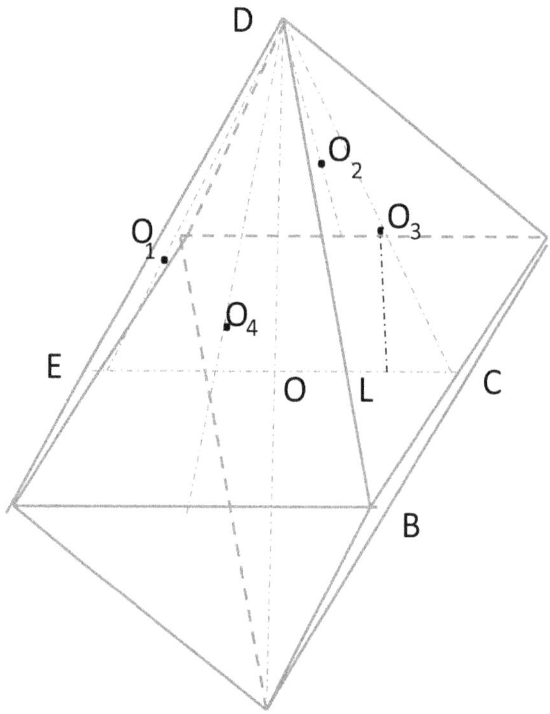

Suppose that in the upper half of an octahedron with an edge the faces of the faces are points O_1, O_2, O_3, O_4. Consider the triangle *EDC* passing through the vertex of the octahedron *D*, the points O_1, O_3 and the midpoints of the two parallel edges of the octahedron of the point *E* and *C*.

Can to find the distance O_1O_3, which is the diagonal of the square that is the face of the inscribed cube. From consideration of two similar triangles O_3LC and *DCO* follows that

$$\frac{LC}{OC} = \frac{CO_3}{DC},$$

but,

$$O_3C = \frac{a}{\sqrt{3}}, OC = \frac{a}{2}, DC = \frac{\sqrt{3}}{2}a.$$

hence,

$$LC = \frac{a}{3}, O_1O_3 = a - 2\frac{a}{3} = \frac{a}{3}.$$

Therefore,

$$O_2O_3 = \frac{a}{3\sqrt{2}}.$$

Thus, at the transition from the cube to the closest - sized octahedron, the length of the edge increases by a factor $3\sqrt{2}$.

If the transition is from an octahedron to a cube, then from the consideration of Figure 8 it follows that the edge of an octahedron (O_1O_2) is the diagonal of an isosceles right triangle with legs $a / 2$ if a is an edge of the described cube.

Figure 8. Hierarchical filling of space with a cube and an octahedron

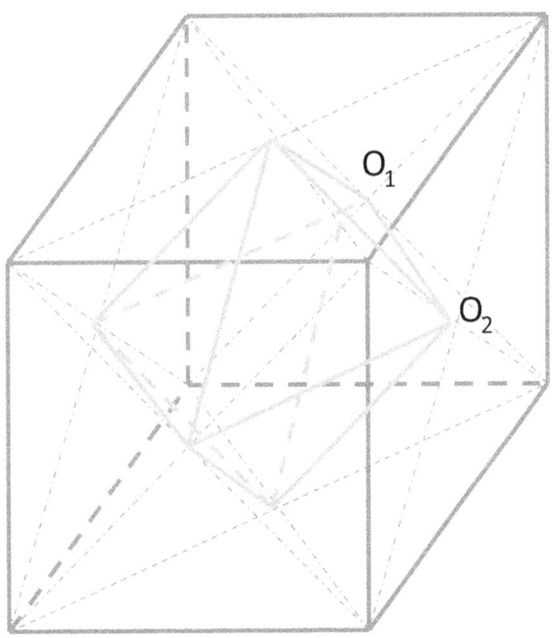

Thus, when a cube is formed from an octahedron, the edge increases by $\sqrt{2}$ times. If we take as the initial figure a cube with an edge 1, then the sequence of edges with alternating cubes and octahedrons looks like Table 2.

From Table 2 it follows that if from this sequence to separate the sequence of cubes and the sequence of octahedrons, then both sequences form a geometric progression with a denominator of $q_{c,oc} = 6$, differing by an initial element of the sequences: for the cube -1, for the octahedron - $3\sqrt{2}$.

Hierarchical Filling of Space With Dodecahedron and Icosahedron

For the dodecahedron and icosahedron, only the second method of hierarchical filling of space is also applicable. Drawing a plane through the vertices of the dodecahedron perpendicular to the diagonals passing through these vertices, can to obtain as a result of the intersections of the planes a closed figure - an icosahedron having 20 faces, as much as a dodecahedron of vertices. This is a consequence of the duality property of the dodecahedron and icosahedron. If one draw the planes through the vertices of the icosahedron, perpendicular to the diagonals of the icosahedron passing through these vertices, then can to get as a result a dodecahedron having 12 faces, as much as a dodecahedron of vertices. Thus, in this case the sequence of polyhedrons consists in the alternation of dodecahedrons and icosahedra: a dodecahedron gives rise to an icosahedron, an icosahedron gives rise to a dodecahedron.

Let the icosahedron be inscribed in the dodecahedron (Figure 9).

Then the middle of the regular pentagons, which are the faces of the dodecahedron, are connected by the sides of one regular triangle, as one of the faces of the icosahedron. To determine the ratio of the sides of a pentagon

Table 2. The length of edges a cube and an octahedron at hierarchical filling of space

———	C	Oc	C	Oc	C	Oc	C	oc
length of edges	1	$3\sqrt{2}$	6	$18\sqrt{2}$	36	$108\sqrt{2}$	216	$648\sqrt{2}$

Figure 9. Hierarchical filling of space with dodecahedron and icosahedron

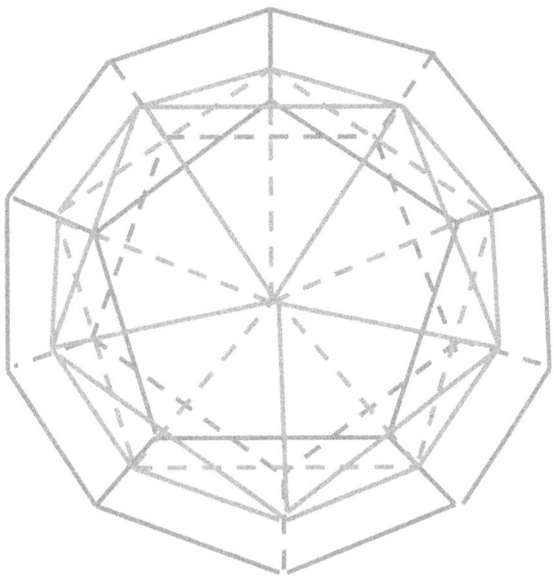

and a triangle, calculate the dihedral angle in the triangular corner formed by the faces of the dodecahedron. Between the plane angles of the triangular angle *abc* and the dihedral angle *A* there is a relation of the spherical geometry (Zalgaller, 1967; Zhizhin, 2009)

$$\cos a = \cos b \cos c + \sin b \sin c \cos A, \tag{14}$$

where the dihedral angle *A* lies opposite the angle *a*. Since in the regular pentagon all angles at the vertices are 108°, hence $a = b = c = 108°$, then from (14) can to find the cosine of the dihedral angle between the faces of the dodecahedron

$$\cos A = \frac{\left(\cos 108^o\right)\left(1 - \cos 108^o\right)}{\left(\sin 108^o\right)^2}. \tag{15}$$

From (15) can to find that $A = 116{,}5^o$.

Consider the triangle *BOL* formed by the edge of the icosahedron and the perpendiculars *OL* and *OB* dropped from the centers of the faces of the dodecahedron to the edge of their connecting (see Figure 10).

Figure 10. The dihedral angle in the dodecahedron

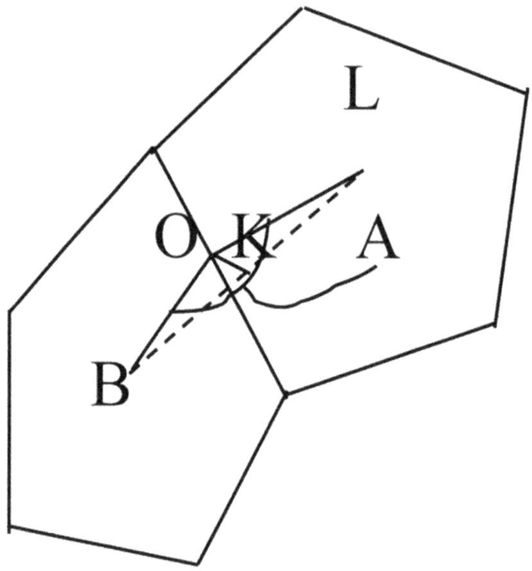

Since $BO = OL = a / 2\ tg\ 54^o$, where a is the length of the edge of the dodecahedron, edges of an inscribed icosahedron

$$2KL = a \sin \frac{A}{2} tg54^o = a1,170409428 \tag{16}$$

It follows from (16) that if the edge of the inscribed icosahedron is taken as 1, then the edge of the described dodecahedron is equal to 0.854401866, that is, the edge of the described dodecahedron is less than the edge of the inscribed icosahedron!

Now let the dodecahedron be inscribed in the icosahedron. Then the middle of the regular triangles forming the pentagonal corner, at the top of the icosahedron, are connected by the sides of the regular pentagon, like the faces of a dodecahedron. Moreover, all dihedral angles in the pentagonal angle are equal (pentagonal angle is completely symmetric). From the relations connecting dihedral angles to the pentagonal angle (Zhizhin, 2009) it follows that the dihedral angles in a completely symmetric pentagonal corner with faces from regular triangles are $138.2^o = \alpha$.

Figure 11. The dihedral angle in the icosahedron

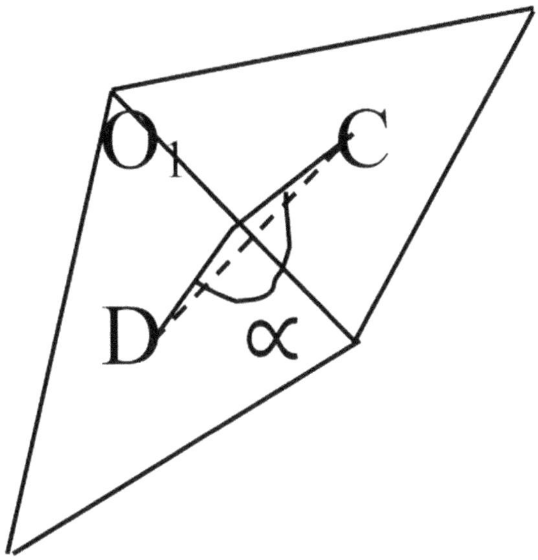

One consider the triangle DO_1C formed by the edge of the dodecahedron DC and the heights O_1C and O_1D from the centers of the faces of the icosahedron (Figure 11).

Since, $O_1C = O_1D = \frac{a}{2\sqrt{3}}$, where a is the edge of the icosahedron, the length of the edge of the inscribed dodecahedron

$$DC = 2\frac{a}{2\sqrt{3}}\frac{1+\sqrt{5}}{2\sqrt{3}} = a\frac{1+\sqrt{5}}{6}.$$

It follows from the last equality that if we take the length of an inscribed dodecahedron as 1, then the length of the edge of the described icosahedron is equal to $\frac{6}{1+\sqrt{5}} = 1,854038233.$

Thus, the hierarchical filling of space with the dodecahedron and icosahedron leads to a sequence of alternating dodecahedrons and icosahedrons.

The lengths of the edges of polyhedrons and their volumes with increasing number of steps "*k*" in the sequence are presented in Table 3, provided that the initial figure ($k = 0$) is a dodecahedron with a length of edge 1.

From Table 3 it follows that, if from the sequence of alternating dodecahedrons and icosahedron isolate the sequences of dodecahedron edges and separately the sequence of edges of icosahedron, then these two sequences form geometric progressions with the denominator $q_{d,i}$ = 1.584093726. However, these sequences have different initial edge values: for the dodecahedron edges sequences, the initial length of the edge 1, for the sequence of edges of icosahedron the initial length of the edge is 1.854038233.

It should be noted that, despite the decrease in the length of the edge, the volume of the described polyhedron is larger than the volume of the inscribed polyhedron from the icosahedron to the described dodecahedron for any generation (see Table 3).

Hierarchical Filling of a Three - Dimensional Space by Convex Semi - Regular Polyhedrons

Among semi - regular polyhedrons (Zhizhin, 2014) there is no dual to itself (the numbers of vertices and faces are different) and there are no pairs of dual polyhedrons. Therefore, semiregular polyhedrons cannot hierarchically fill the space in a way with common points that increase their size by polyhedron. Semi - regular polyhedrons can hierarchically fill space only in the second way, i.e. not having common points with each other.

Variants of the hierarchical filling of space by semi - regular convex three - dimensional polytopes have been studied successively in Zhizhin's book

Table 3. The length of the edges of dodecahedron and icosahedron at hierarchical filling of space

k	0	1	2	3	4	5
a dodecahedron	1		1,584093726		2,509352933	
a icosahedron		1,854038233		2,93697033		4,652436278
V dodecahedron	7,66		30,44888481		121,0358468	
V Icosahedron		13,89352832		55,22747305		219,5319799

(Zhizhin, 2014). The similarity coefficients for all semi - regular bodies are determined. More precisely, these similarity coefficients should be called self - similar coefficients. The values of the self - similarity coefficients determine the regularities of the hierarchical filling of space by convex semi - regular polyhedrons. It is essential that semi -regular polyhedrons, in contrast to regular polyhedrons, have as a rule several (from 1 to 5) similarity coefficients, i.e. several variants of hierarchical filling of space by each polyhedron. This leads to the possibility of simultaneously increasing the volume in several ways at each step of the filling, which considerably complicates the filling structure and contributes to the densest filling of the space. In addition, this leads to the possibility of a hierarchical chaotic filling of the space, since at each step of filling different ways of filling can be arbitrarily realized.

HIERARCHICAL FILLING OF HIGHER DIMENSIONAL SPACE

Now, applying the methods developed in previous studies, one to consider the hierarchical filling of a multidimensional space in the process of enlargement of multidimensional convex bodies (Zhizhin, 2012). As was shown earlier when studying the hierarchical filling of a plane with polygons (Zhizhin, 2010), there can be two ways of filling it. In a multidimensional space is first way of hierarchical filling with polytopes, i.e. continuation of the edges of the figure previous stage before intersecting them outside the figure, forming the vertices of the figure in a new stage. For the three - dimensional polytope the method of hierarchical filling of space was based on the transition of a polytope to a dual polytope at each step of filling the space. In this case, investigating scaling processes, one need to consider a method for filling a space of higher dimension with polytopes, at which only the scale of the figure would change at each step, and the polytope's background would be preserved.

On Figure 4 of Chapter 3 presented projection of the 4 – simplex.

Continuing the outer edges of the projection to the intersection with each other, connecting the resulting vertices with edges, can to obtain an enlarged projection of the 4 - simplex (Figure 12).

It should be emphasized that the lines that extend the edges of the figure are of an auxiliary character (however, as in the hierarchical filling of the plane). In this case, the edges of the figure of the next generation, due to the

Figure 12. The hierarchical filling of the space by a 4 - simplex

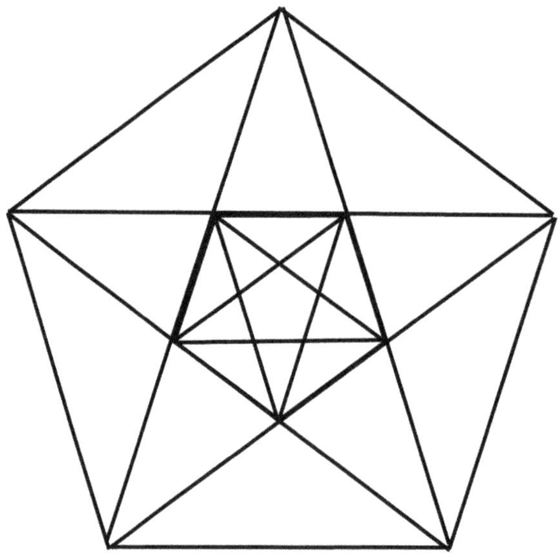

multidimensionality of space in the projection, can be partially superimposed on the auxiliary lines and edges of the figures of previous generations.

Obviously, the process of extending the 4 - simplex can be continued further. Each time the size (length) of the edges of the next 4 - simplex from the size of the edges of the previous 4 - simplex is increased by $1 + \tau$ times. Where τ is the golden section. Since the coefficient of the geometric progression of the discrete increase of each side of the regular pentagon in the hierarchical filling of the pentagon plane is $1 + \tau$ (Zhizhin, 2010), then each segment joining the vertices of the pentagon increases in the same progression. The difference in the lengths of the segments of the projection of a regular the 4 - simplex into a two - dimensional plane is related to the design process, but their images in four - dimensional space have the same length. Thus, the geometric progression denominator of the hierarchical filling of the space by a 4 - simplex is $1 + \tau$.

On Figure 10 of Chapter 3 presented known projection of the 4 – cube in symmetrical form.

Continuing the outer edges of the projection to the intersection with each other, connecting the resulting vertices with edges, can to obtain an enlarged projection of the 4 - cube (Figure 13).

Obviously, the process of extending the 4 - cube can be continued further. Easy to see that each time the size (length) of the edges of the next 4 - cube

Figure 13. The hierarchical filling of the space by a 4 - cube

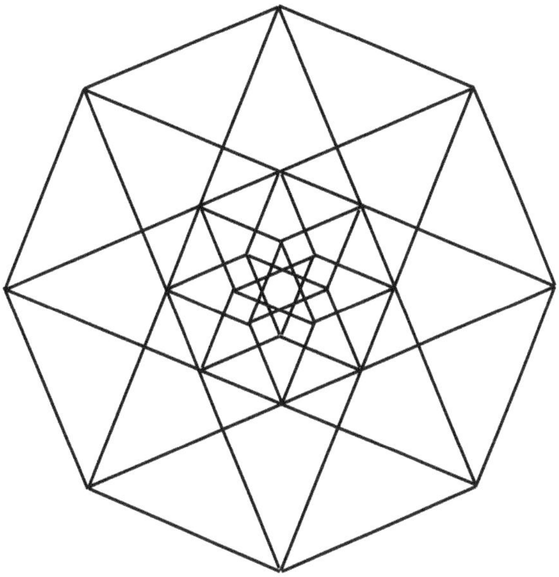

Figure 14. The hierarchical filling of the space by a 4 – cross - polytope

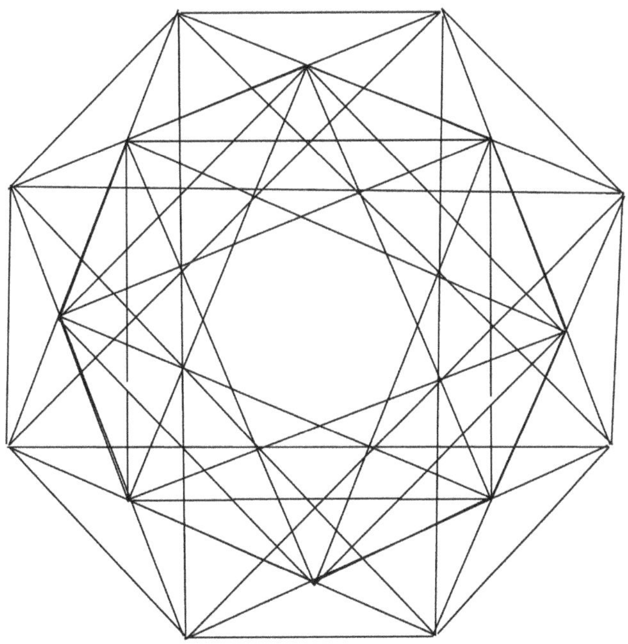

from the size of the edges of the previous 4 - cube is increased by $1+\sqrt{2}$ times. Thus, the geometric progression coefficient of the hierarchical filling of the space by a 4 - cube is $1+\sqrt{2}$.

On Figure 9 of Chapter 2 presented projection of the 4 –cross - polytope in symmetrical form.

Continuing the outer edges of the projection to the intersection with each other, connecting the resulting vertices with edges, can to obtain an enlarged projection of the 4 - cross - polytope (Figure 14).

Obviously, the process of extending the 4 – cross - polytope can be continued further. Easy to see that each time the size (length) of the edges of the next 4 – cross - polytope from the size of the edges of the previous 4 – cross - polytope is increased by $\sqrt{1+\frac{1}{\sqrt{2}}}$ times. Thus, the geometric progression denominator of the hierarchical filling of the space by a 4 – cross - polytope is $\sqrt{1+\frac{1}{\sqrt{2}}}$.

Figure 15. The hierarchical filling of the space by a 5 - simplex

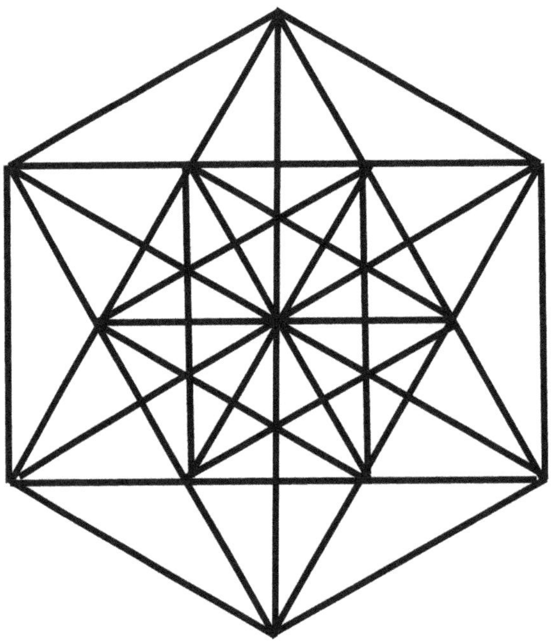

In the same way, it is possible to construct a hierarchical filling of a space of dimension greater than four starting from some initial figure of this dimension. For example, let this figure be the 5 - simplex. Its image on the plane is shown in Figure 2 of Chapter 3.

This polytope has 6 of the vertices, 16 of the edges, 20 of the two - dimensional faces, 15 of the three – dimensional faces, 6 of the four - dimensional faces. Continuing the outer edges of the projection to the intersection with each other, connecting the resulting vertices with edges, can to obtain an enlarged projection of the 5 - simplex (Figure 15).

Obviously, the process of extending the 5 - simplex can be continued further. Easy to see that each time the size (length) of the edges of the next 5 - simplex from the size of the edges of the previous 5 - simplex is increased by $\sqrt{3}$ times. Thus, the geometric progression denominator of the hierarchical filling of the space by a 4 - simplex is $\sqrt{3}$.

CONCLUSION

The process of scaling (zooming in) has attracted the attention of researchers in recent years in analyzing various phenomena. In this chapter, we study the geometric characteristics of the scaling process in the hierarchical filling of spaces. The investigation was started from the hierarchical filling of the plane by regular polygons in the "growing geometric variety" regime. This mode simulates the growth of biological objects, for example, flowers. Two ways of the scaling process are discovered, which also manifest themselves in spaces of higher dimension. The denominator of geometric progressions characterizing a discrete series of polygon sizes in the process of hierarchical filling of the plane are determined. These denominators can be regarded as the values of the scaling function of the corresponding processes. It can be noted that changing the size of the figures during the hierarchical filling of the plane does not obey the Fibonachchi series for any number of sides of the polygon. The hierarchical filling of a three -dimensional space by regular and semi - regular polyhedrons is investigated. In the hierarchical filling of space by three - dimensional polyhedrons at each level of filling, the type of the polyhedron changes by a dual type. When the space is filled with semi - regular polyhedrons, each variant of a semi - regular polyhedron can have different variants of changing the size of the figure. This leads to possible chaotic hierarchical filling of space. The value of its scaling function is calculated for each type of three - dimensional polyhedron. The processes of hierarchical filling of a space of higher dimension by convex polytopes are also considered. These fillings are constructed for a number of polytopes of higher dimension, and the values of their scaling functions are determined.

REFERENCES

Barenblatt, G. I. (1978). *Similarity, self - similarity, intermediate asymptotes.* Leningrad: Hydrometioizdat.

Grünbaum, B & Shephard, G. (1987). *Tilings and Patterns.* New York: A series of Books in the Mathematical Sciences.

Keppleri, J. (1611). *Strena, seu de nive sexangula.* Tampach: Francofurti ad Moenum.

Lord, E. A., Mackay, A. L., & Ranganathan, S. (2006). *New Geometry for New Materials.* Cambridge: Cambridge University Press.

Mackay, A. L. (1981). De Nive Quinquangula: On the pentagonal snowflake. *Kristallografiya., 26*, 910–919.

Novak, F. A. (1976). *Illustrated encyclopedia of plants.* Prague: Artia.

Penrose, R. (1974). The role of aesthetics in pure and applied mathematical research. *Bulletin - Institute of Mathematics and Its Applications, 10*, 266.

Prochorov, A. M. (Ed.). (1983). *Physical encyclopedic dictionary.* Moscow: Soviet Encyclopedia.

Takhtadzhyan, A. L. (Ed.). (1980). *Life of plants. T.5. Flowering plants.* Moscow: Enlightenment.

Vladimirov, Yu. (2002). *Metaphysics.* Moscow: Binom. Knowledge lab.

Vorobiev, N. N. (1984). *Fibonacci numbers.* Moscow: Science.

Zalgaller, V. A. (1966). *Convex polyhedrons with correct sides.* Leningrad: Science.

Zhizhin, G. V. (1993). *Parquetages of equal triangles, adjoining on whole sides.* Saint - Petersburg: Polytechnika.

Zhizhin, G. V. (2009). *Convex Right – Faceted Polyhedrons*. St. Petersburg: Polytechnika – service.

Zhizhin, G. V. (2010). *Geometric bases of the dissipative structures*. Saint – Petersburg: Polytechnika.

Zhizhin, G. V. (2012, October). Hierarchical filling of spaces with polytopes. *Paper presented at 7th St.- Petersburg meeting of Nobel Prize laureates*, St. Petersburg, Russia.

Zhizhin, G. V. (2014a). *World – 4D*. St. Petersburg: Polytechnic Service.

Zhizhin, G. V. (2017). *Chemical Compound Structures and the Higher Dimension of Molecules: Emerging Research and Opportunities*. Hershey, PA: IGI Global.

KEY TERMS AND DIFINITIONS

Denominator of Geometric Progression: The relationship between the previous and subsequent elements of a geometric progression.

Hierarchical Filling of Space: The filling of space by a figure with its discrete resizing, preserving the resemblance of figures at each step of its change.

Scaling: Scale change of shape.

Chapter 8
On the Possible Electronic Structure of Atoms in a Space of Higher Dimension

ABSTRACT

It is shown that the stationary Schrödinger equation describing the distribution of electrons in the vicinity of the atomic nucleus has a solution, in principle, for any dimensionality of the space around the nucleus. As an example, a solution of the Schrödinger equation in a five - dimensional space is obtained. It is shown that the solution of the Schrödinger equation in p - dimensional space has p quantum numbers: the principal quantum number, the orbital quantum number and p - 2 magnetic quantum numbers. Taking into account the spin quantum number, the total number of quantum numbers in p - dimensional space is p + 1. This leads to the possibility of increasing the number of quantum cells of orbitals and, consequently, to the possibility of increasing the valence of the elements.

INTRODUCTION

The studies carried out in the monograph of G. Zhizhin "Chemical Compound Structures and Higher Dimension of Molecules: Emerging Research and Opportunities" (Zhizhin, 2018) convincingly showed that many elements of the periodic system of Mendeleev elements form molecules whose spatial structure can be geometrically represented as a convex polytope. Moreover,

DOI: 10.4018/978-1-5225-6968-8.ch008

the dimension of this polytope, determined by the Euler-Poincaré relation between the geometric elements forming the polytope (Poincare, 1895), is usually higher than three. In addition, in this monograph it is noted that most elements of the periodic system exhibit valence in compounds higher than the valence determined by the location of the element in the periodic table. It is known that the calculation of the valence of a chemical element based on its location in the periodic table is based on the electronic structure of atoms (s -, p -, d -, f – orbitals), determined by solving the Schrodinger equation (Schrodinger, 1927) under the assumption of the three - dimensionality of the space surrounding the nucleus of the atom. To explain the observed differences in the valence of atoms from valence, determined from the location of the element in the periodic table, researchers resorted to the construction of various models of the interaction of electrons in the vicinity of atomic nuclei. These models are quite diverse and still remain at the level of hypotheses. In this connection, this chapter analyzes the possibility of obtaining a solution of the Schrödinger equation in a space of higher dimension. Although before it was believed that the solutions of the Schrödinger equation in a space with a dimension greater than three do not exist (Büchel, 1963; Freeman, 1969; Gurevich & Mostepanenko, 1971).

THE STATIONARY SCHRÖDINGER EQUATION IN A *P* - DIMENSIONAL METRIC SPACE

If a solution of the Schrödinger equation in the case of four - dimensional space was obtained in the monograph of Zhizhin (Zhizhin, 2017), in this book the solution of the Schrodinger equation for any finite dimensionality of space is obtained in principle.

It is shown that the solution of the Schrödinger equation in a space of higher dimension leads to the existence of a large number of quantum numbers and, consequently, to an increase in the number of quantum cells of atoms and an increase in their valence.

The stationary Schrödinger equation in a p - dimensional metric space has view

$$-\frac{\hbar}{2m}\nabla_p^2\Psi + U\Psi = E\Psi. \qquad (1)$$

where Ψ - the wave function, U - the potential energy of an electron (the function of radius *r*), *E* – the kinetic energy of electron, ∇_p^2 - Laplacian in *p* - dimensional metric space, *m* – the mass of electron, $\hbar$ - Planck constant.

The potential energy *U(r)* – this potential energy of the Coulomb attraction of an electron to the nucleus and can be written in explicit form

$$U(r) = -\frac{Ze^2}{r} \tag{2}$$

where eZ - ion's charge, *e* – electron's charge.

It is essential that the form of the potential energy (2) does not depend on the dimensionality of space and it is defined as a function inversely proportional to the radius in the first degree. Since in a space of any dimension, the potential energy is a function of the distance between the charges (of the electron and the nucleus). The incorrectness in previous studies of the solution of the Schrödinger equation in spaces of higher dimension is that the authors of these works (Büchel, 1963; Freeman, 1969; Gurevich & Mostepanenko, 1971) recorded potential energy without a physical basis as a function of the radius to the power of *p* -2. Such an entry goes to formula (2) only in three - dimensional space, but in a space of the high dimensional this record is not true and leads to an absurd conclusion about the impossibility of the existence of a discrete atomic structure in a space of higher dimension.

Equation (1), having transferred to the atomic system of units, taking into account (2), in the Cartesian coordinate system has the form

$$\frac{\partial^2\Psi}{\partial x_1^2} + \frac{\partial^2\Psi}{\partial x_2^2} + \ldots + \frac{\partial^2\Psi}{\partial x_p^2} + \left(\frac{c}{r} + 2E\right)\Psi = 0. \tag{3}$$

where c – constant, $r = \sqrt{x_1^2 + \ldots + x_p^2}$.

For further analysis, it is convenient to go over into a spherical coordinate system, where the radius of one of the *p* variables. The remaining *p* - 1 variables are the angles between the vector of length *r* to the point of location of the particle and the unit vectors of the Cartesian system of coordinates.

It can be noted that Gurevich & Mostepanenko (1971) were confused with coordinate systems. For example, the velocity vector in the *n* - dimensional space has *n* + 1 components in their work. Babenko (2015), considering *n*

- dimensional space, describes the one - dimensional Schrödinger equation, forgetting the remaining variables.

In an n - dimensional space, the analytic connection between the coordinates is expressed by equalities (Hobson, 1931)

$$\begin{aligned} x_1 &= r \sin \alpha_1 \sin \alpha_2 \ldots \sin \alpha_{p-1}, \\ x_2 &= r \cos \alpha_1 \sin \alpha_2 \ldots \sin \alpha_{p-1}, \\ x_3 &= r \cos \alpha_2 \sin \alpha_3 \ldots \sin \alpha_{p-1}, \\ &. \\ &. \\ &. \\ x_{p-1} &= r \cos \alpha_{p-2} \sin \alpha_{p-2}, \\ x_p &= r \cos \alpha_{p-1} \end{aligned} \tag{4}$$

The Laplacian in the Schrodinger equation in a spherical coordinate system in p - dimensional space can be written in the form

$$\nabla_p^2 = \frac{1}{L_1 L_2 L_3 \ldots L_p} \left[\frac{\partial}{\partial r} \left(\frac{L_2 L_3 \ldots L_p}{L_1} \frac{\partial \Psi}{\partial r} \right) + \frac{\partial}{\partial \alpha_1} \left(\frac{L_1 L_3 \ldots L_p}{L_2} \frac{\partial \Psi}{\partial \alpha_1} \right) + \frac{\partial}{\partial \alpha_2} \left(\frac{L_1 L_2 L_4 \ldots L_p}{L_3} \frac{\partial \Psi}{\partial \alpha_2} \right) + \right.$$

$$\left. + \frac{\partial}{\partial \alpha_3} \left(\frac{L_1 L_2 L_3 L_5 \ldots L_p}{L_4} \frac{\partial \Psi}{\partial \alpha_3} \right) + \ldots + \frac{\partial}{\partial \alpha_{p-1}} \left(\frac{L_1 L_2 \ldots L_{p-1}}{L_p} \frac{\partial \Psi}{\partial \alpha_{p-1}} \right) \right]. \tag{5}$$

where L_i - the Lame coefficients, $i = 1, 2, \ldots, p$,

$$L_1 = \sqrt{\left(\frac{\partial x_1}{\partial r}\right)^2 + \left(\frac{\partial x_2}{\partial r}\right)^2 + \ldots + \left(\frac{x_p}{\partial r}\right)^2},$$

$$L_2 = \sqrt{\left(\frac{\partial x_1}{\partial \alpha_1}\right)^2 + \left(\frac{\partial x_2}{\partial \alpha_1}\right)^2 + \ldots + \left(\frac{x_p}{\partial \alpha_1}\right)^2},$$

$$\cdot \quad \cdot \quad \cdot \tag{6}$$

$$L_p = \sqrt{\left(\frac{\partial x_1}{\partial \alpha_{p-1}}\right)^2 + \left(\frac{\partial x_2}{\partial \alpha_{p-1}}\right)^2 + \ldots + \left(\frac{x_p}{\partial \alpha_{p-1}}\right)^2}$$

Differentiating the equality (4) with respect to the independent variables, substituting the derivatives in the Lame coefficients (6), can to obtain after the transformations

$$\begin{aligned} &L_1 = 1, L_2 = r \sin \alpha_2 \ldots \sin \alpha_{p-1}, L_3 = r \sin \alpha_3 \ldots \sin \alpha_{p-1}, \\ &L_4 = r \sin \alpha_4 \ldots \sin \alpha_{p-1}, \ldots, L_{p-1} = r \sin \alpha_{p-1}, L_p = r. \end{aligned} \tag{7}$$

The Schrödinger equation in a spherical coordinate system in p - dimensional space has view

$$\nabla_p^2 \Psi + \left(\frac{2}{r} + 2E\right)\Psi = 0. \tag{8}$$

where $\nabla_p^2 \Psi$ is expressed using equations (5) – (7).

Than Schrödinger equation (8) has view

$$\begin{aligned} &\frac{1}{r^{p-1}} \frac{\partial}{\partial r}\left(r^{p-1} \frac{\partial \Psi}{\partial r}\right) + \frac{1}{r^2 \sin \alpha_2{}^2 \ldots \sin \alpha_{p-1}{}^2} \frac{\partial^2 \Psi}{\partial \alpha_1{}^2} + \frac{1}{r^2 \sin \alpha_2 \sin \alpha_3{}^2 \ldots \sin^2 \alpha_{p-1}} \left(\sin \alpha_2{}^2 \frac{\partial \Psi}{\partial \alpha_2}\right) + \\ &\frac{1}{r^2 \sin \alpha_3{}^2 \sin^2 \alpha_4 \ldots \sin^2 \alpha_{p-1}} \frac{\partial}{\partial \alpha_3}\left(\sin \alpha_3{}^2 \frac{\partial \Psi}{\partial \alpha_3}\right) + \ldots + \frac{1}{r^2 \sin^2 \alpha_{p-1}} \frac{\partial}{\partial \alpha_{p-1}}\left(\sin^2 \alpha_{p-1} \frac{\partial \Psi}{\partial \alpha_{p-1}}\right) + \\ &+2\frac{\Psi}{r} + 2E\Psi = 0. \end{aligned} \tag{9}$$

THE DECISION OF THE STATIONARY SCHRÖDINGER EQUATION IN A P - DIMENSISION METRIC SPACE

A solution of equation (9) can be obtained by separating the variables assuming that

$$\Psi\left(r,\alpha_1,\ldots,\alpha_{p-1}\right)=R\left(r\right)\Phi_1\left(\alpha_1\right)\Phi_2\left(\alpha_2\right)\ldots\Phi_{p-1}\left(\alpha_{p-1}\right). \tag{10}$$

We substitute (10) into (9)

$$\frac{1}{r^{p-3}R(r)}\frac{d}{dr}\left(r^{p-1}\frac{dR(r)}{dr}\right)+2r^2\left(\frac{1}{r}+E\right)=-\frac{1}{\Phi_1(\alpha_1)\sin\alpha_2{}^2\ldots\sin\alpha_{p-1}{}^2}\frac{d^2\Phi_1(\alpha_1)}{d\alpha_1{}^2}-$$
$$-\frac{1}{\Phi_2(\alpha_2)\sin\alpha_2\sin\alpha_3{}^2\ldots\sin^2\alpha_{p-1}}\frac{d}{d\alpha_2}\left(\sin\alpha_2\frac{d\Phi_2(\alpha_2)}{d\alpha_2}\right)-$$
$$-\frac{1}{\Phi_3(\alpha_3)\sin\alpha_3{}^2\sin^2\alpha_4\ldots\sin^2\alpha_{p-1}}\frac{d}{d\alpha_3}\left(\sin\alpha_3{}^2\frac{d\Phi_3(\alpha_3)}{d\alpha_3}\right)-$$
$$-\ldots-\frac{1}{\Phi_{p-1}\sin^2\alpha_{p-1}}\frac{d}{d\alpha_{p-1}}\left(\sin^2\alpha_{p-1}\frac{d\Phi_{p-1}}{d\alpha_{p-1}}\right). \tag{11}$$

The left part of (11) is depended at radius *r*, and rite part is dependent only at angels.

Therefore, both parts equality (11) equal to same constant C_1. From left part of (11) there is

$$\frac{1}{r^{p-3}R(r)}\frac{d}{dr}\left(r^{p-1}\frac{dR(r)}{dr}\right)+2r^2\left(\frac{1}{r}+E\right)=C_1. \tag{12}$$

From right part of (11) there is

$$C_1 = -\frac{1}{\Phi_1(\alpha_1)\sin\alpha_2^{\;2}...\sin\alpha_{p-1}^{\;2}}\frac{d^2\Phi_1(\alpha_1)}{d\alpha_1^{\;2}} -$$
$$-\frac{1}{\Phi_2(\alpha_2)\sin\alpha_2\sin\alpha_3^{\;2}...\sin^2\alpha_{p-1}}\frac{d}{d\alpha_2}\left(\sin\alpha_2\frac{d\Phi_2(\alpha_2)}{d\alpha_2}\right) -$$
$$-\frac{1}{\Phi_3(\alpha_3)\sin\alpha_3^{\;2}...\sin^2_{p-1}}\frac{d}{d\alpha_3}\left(\sin\alpha_3^{\;2}\frac{d\Phi_3(\alpha_3)}{d\alpha_3}\right) - ... - \frac{1}{\Phi_{p-1}\sin^2\alpha_{p-1}}\frac{d}{d\alpha_{p-1}}\left(\sin^2\alpha_{p-1}\frac{d\Phi_{p-1}}{d\alpha_{p-1}}\right) \quad (13)$$

The equation (13) can to rewrite as follows

$$C_1\sin\alpha_{p-1}^{\;2} + \frac{1}{\Phi_{p-1}(\alpha_{p-1})}\frac{d}{d\alpha_{p-1}}\left(\sin\alpha_{p-1}^{\;2}\frac{d\Phi_{p-1}(\alpha_{p-1})}{d\alpha_{p-1}}\right) = -\frac{1}{\Phi_1(\alpha_1)\sin\alpha_2^{\;2}...\sin^2_{p-2}}\frac{d^2\Phi_1(\alpha_1)}{d\alpha_1^{\;2}} -$$
$$-\frac{1}{\Phi_2(\alpha_2)\sin\alpha_2\sin^2\alpha_3...\sin^2\alpha_{p-2}}\frac{d}{d\alpha_2}\left(\sin\alpha_2\frac{d\Phi_2(\alpha_2)}{d\alpha_2}\right) -$$
$$-\frac{1}{\Phi_3\left(\alpha_3\right)\sin^2\alpha_3...\sin^2\alpha_{p-2}}\frac{d}{d\alpha_3}\left(\sin^2\alpha_3\frac{d\Phi_3\left(\alpha_3\right)}{d\alpha_3}\right) - ...$$
$$-\frac{1}{\Phi_{p-2}\left(\alpha_{p-2}\right)\sin^2\alpha_{p-2}}\frac{d}{d\alpha_{p-2}}\left(\sin^2\alpha_{p-2}\frac{d\Phi_{p-2}\left(\alpha_{p-2}\right)}{d\alpha_{p-2}}\right). \quad (14)$$

The left part of (14) is depended at angle α_{p-1} only, and the right part (14) depended at angles $\alpha_2,...,\alpha_{p-2}$. Therefore, both parts of (14) equal to same constant C_2. From the left part (14) there is

$$C_1\sin\alpha_{p-1}^{\;2} + \frac{1}{\Phi_{p-1}(\alpha_{p-1})}\frac{d}{d\alpha_{p-1}}\left(\sin\alpha_{p-1}^{\;2}\frac{d\Phi_{p-1}(\alpha_{p-1})}{d\alpha_{p-1}}\right) = C_2 \quad (15)$$

From the right part (14) there is

$$-\frac{1}{\Phi_1(\alpha_1)}\frac{d^2\Phi_1(\alpha_1)}{d\alpha_1^{\ 2}} = C_2 \sin^2\alpha_2 ... \sin^2\alpha_{p-2} + \frac{\sin\alpha_2}{\Phi_2(\alpha_2)}\frac{d}{d\alpha_2}\left(\sin\alpha_2 \frac{d\Phi_2(\alpha_2)}{d\alpha_2}\right) + \\ + \frac{\sin^2\alpha_2}{\Phi_3(\alpha_3)}\frac{d}{d\alpha_3}\left(\sin^2\alpha_3 \frac{d\Phi_3(\alpha_3)}{d\alpha_3}\right) + ... + \frac{\sin^2\alpha_2 ... \sin^2\alpha_{p-3}}{\Phi_{p-2}(\alpha_{p-2})}\frac{d}{d\alpha_{p-2}}\left(\sin^2\alpha_{p-2}\frac{d\Phi_{p-2}(\alpha_{p-2})}{d\alpha_{p-2}}\right) \tag{16}$$

The left part of (16) is depended at angle α_1 only, and the right part (16) depended at angles $\alpha_2,...,\alpha_{p-2}$. Therefore, both parts of (16) equal to same constant C_3. From the left part (16) there is

$$-\frac{1}{\Phi_1(\alpha_1)}\frac{d^2\Phi_1(\alpha_1)}{d\alpha_1^{\ 2}} = C_3 \tag{17}$$

From the right part (16) there is

$$C_3 = C_2 \sin^2\alpha_2 ... \sin^2\alpha_{p-2} + \frac{\sin\alpha_2}{\Phi_2(\alpha_2)}\frac{d}{d\alpha_2}\left(\sin\alpha_2 \frac{d\Phi_2(\alpha_2)}{d\alpha_2}\right) + \\ + \frac{\sin^2\alpha_2}{\Phi_3(\alpha_3)}\frac{d}{d\alpha_3}\left(\sin^2\alpha_3 \frac{d\Phi_3(\alpha_3)}{d\alpha_3}\right) + ... + \frac{\sin^2\alpha_2 ... \sin^2\alpha_{p-3}}{\Phi_{p-2}(\alpha_{p-2})}\frac{d}{d\alpha_{p-2}}\left(\sin^2\alpha_{p-2}\frac{d\Phi_{p-2}(\alpha_{p-2})}{d\alpha_{p-2}}\right) \tag{18}$$

The equation (18) can to rewrite as follows

$$\frac{C_3}{\sin^2\alpha_2} - \frac{1}{\Phi_2(\alpha_2)\sin\alpha_2}\frac{d}{d\alpha_2}\left(\sin\alpha_2 \frac{d\Phi_2(\alpha_2)}{d\alpha_2}\right) = C_2 \sin^2\alpha_3 ... \sin^2\alpha_{p-2} + \\ + \frac{1}{\Phi_3(\alpha_3)}\frac{d}{d\alpha_3}\left(\sin^2\alpha_3 \frac{d\Phi_3(\alpha_3)}{d\alpha_3}\right) + ... + \frac{\sin^2\alpha_3 ... \sin^2\alpha_{p-3}}{\Phi_{p-2}(\alpha_{p-2})}\frac{d}{d\alpha_{p-2}}\left(\sin^2\alpha_{p-2}\frac{d\Phi_{p-2}(\alpha_{p-2})}{d\alpha_{p-2}}\right) \tag{19}$$

The left part of (19) is depended at angle α_2 only, and the right part (19) depended at angles $\alpha_3,...,\alpha_{p-2}$. Therefore, both parts of (19) equal to same constant C_4. From the left part (19) there is

$$\frac{C_3}{\sin^2 \alpha_2} - \frac{1}{\Phi_2(\alpha_2)\sin \alpha_2}\frac{d}{d\alpha_2}\left(\sin \alpha_2 \frac{d\Phi_2(\alpha_2)}{d\alpha_2}\right) = C_4 \tag{20}$$

From the right part (19) there is

$$C_4 = C_2 \sin^2 \alpha_3 ... \sin^2 \alpha_{p-2} + \\ + \frac{1}{\Phi_3(\alpha_3)}\frac{d}{d\alpha_3}\left(\sin^2 \alpha_3 \frac{d\Phi_3(\alpha_3)}{d\alpha_3}\right) + ... + \frac{\sin^2 \alpha_3 ... \sin^2 \alpha_{p-3}}{\Phi_{p-2}(\alpha_{p-2})}\frac{d}{d\alpha_{p-2}}\left(\sin^2 \alpha_{p-2} \frac{d\Phi_{p-2}(\alpha_{p-2})}{d\alpha_{p-2}}\right) \tag{21}$$

The equation (21) can to rewrite as follows

$$\frac{C_4}{\sin^2 \alpha_3} - \frac{1}{\Phi_3(\alpha_3)\sin^2 \alpha_3}\frac{d}{d\alpha_3}\left(\sin^2 \alpha_3 \frac{d\Phi_3(\alpha_3)}{d\alpha_3}\right) = C_2 \sin^2 \alpha_4 ... \sin^2 \alpha_{p-2} + \\ + \frac{1}{\Phi_4(\alpha_4)}\frac{d}{d\alpha_4}\left(\sin^2 \alpha_4 \frac{d\Phi_4(\alpha_4)}{d\alpha_4}\right) + ... + \frac{\sin^2 \alpha_4 ... \sin^2 \alpha_{p-3}}{\Phi_{p-2}(\alpha_{p-2})}\frac{d}{d\alpha_{p-2}}\left(\sin^2 \alpha_{p-2} \frac{d\Phi_{p-2}(\alpha_{p-2})}{d\alpha_{p-2}}\right) \tag{22}$$

From the preceding it is clear that the process of separating equations for unknown angles can be continued to the end, i.e. until a complete system of equations for these unknowns is obtained. This, in principle, allows us to solve a system of equations for any finite value of p; for any finite dimension of the space, in principle, obtain a solution of the Schrödinger equation. Let, for example, the dimension of the space be five ($p = 5$). Then the equation (14) takes the form

$$C_1 \sin \alpha_4^{\ 2} + \frac{1}{\Phi_4(\alpha_4)}\frac{d}{d\alpha_4}\left(\sin \alpha_4^{\ 2} \frac{d\Phi_4(\alpha_4)}{d\alpha_4}\right) = -\frac{1}{\Phi_1(\alpha_1)\sin \alpha_2^{\ 2} \sin^2_{\ 3}}\frac{d^2\Phi_1(\alpha_1)}{d\alpha_1^{\ 2}} - \\ - \frac{1}{\Phi_2(\alpha_2)\sin \alpha_2 \sin^2 \alpha_3}\frac{d}{d\alpha_2}\left(\sin \alpha_2 \frac{d\Phi_2(\alpha_2)}{d\alpha_2}\right) - \frac{1}{\Phi_3(\alpha_3)\sin^2 \alpha_3}\frac{d}{d\alpha_3}\left(\sin^2 \alpha_3 \frac{d\Phi_3(\alpha_3)}{d\alpha_3}\right). \tag{23}$$

Since both parts of the equation (23) depend on different variables, they are equal to some constant. From left part equation (23) there is

$$C_1 \sin{\alpha_4}^2 + \frac{1}{\Phi_4(\alpha_4)} \frac{d}{d\alpha_4}\left(\sin{\alpha_4}^2 \frac{d\Phi_4(\alpha_4)}{d\alpha_4}\right) = C_2. \tag{24}$$

From right part of equation (23) there is

$$\begin{aligned} \frac{1}{\Phi_1(\alpha_1)} \frac{d^2\Phi_1(\alpha_1)}{d{\alpha_1}^2} &= C_2 \sin^2\alpha_2 \sin^2\alpha_3 + \frac{\sin\alpha_2}{\Phi_2(\alpha_2)} \frac{d}{d\alpha_2}\left(\sin\alpha_2 \frac{d\Phi_2(\alpha_2)}{d\alpha_2}\right) + \\ &+ \frac{\sin^2\alpha_2}{\Phi_3(\alpha_3)} \frac{d}{d\alpha_3}\left(\sin^2\alpha_3 \frac{d\Phi_3(\alpha_3)}{d\alpha_3}\right). \end{aligned} \tag{25}$$

Since both parts of the equation (25) depend on different variables, they are equal to some constant. From left part equation (25) can to obtain equation (17).

The equation (17) for any dimension *p* have decision

$$\Phi_1(\alpha_1) = A e^{\pm i\sqrt{C_3}\alpha_1}. \tag{26}$$

Since $\Phi_1(0) = \Phi_1(2\pi)$, then

$$A = A e^{\pm i\sqrt{C_3}\,2\pi}, \tag{27}$$

i.e. $e^{\pm i\sqrt{C_3}\,2\pi} = 1.$

Consequently, $\cos\left(2\pi\sqrt{C_3}\right) \pm i\sin\left(2\pi\sqrt{C_3}\right) = 1.$. It may be only if $\sqrt{C_3} = 0, \pm1, \pm2, \ldots$.

So we got the first quantum number

$$m_1 = \sqrt{C_3} = 0, \pm1, \pm2, \ldots \tag{28}$$

The constant *A* is defined from condition of normalization

$$\int_0^{2\pi} \Phi_1^*(\alpha_1)\Phi_1(\alpha_1)\,d\alpha_1 = 1 = A^2 \int_0^{2\pi} e^{-i\sqrt{C_3}\alpha_1} e^{i\sqrt{C_3}\alpha_1}\,d\alpha_1 = A^2 2\pi.$$

Consequently, $A = \frac{1}{\sqrt{2\pi}}$. Thus, the function $\Phi_1\left(\alpha_1\right)$ it is known for any

p

$$\Phi_1\left(\alpha_1\right) = \frac{1}{\sqrt{2\pi}} e^{-im_1\alpha_1}, m_1 = \sqrt{C_3}. \tag{29}$$

From left part equation (25) there is

$$\begin{aligned} &\frac{C_3}{\sin^2\alpha_2} - \frac{1}{\Phi_2(\alpha_2)\sin\alpha_2}\frac{d}{d\alpha_2}\left(\sin\alpha_2\frac{d\Phi_2(\alpha_2)}{d\alpha_2}\right) = C_2\sin^2\alpha_3 + \\ &+\frac{1}{\Phi_3\left(\alpha_3\right)}\frac{d}{d\alpha_3}\left(\sin^2\alpha_3\frac{d\Phi_3\left(\alpha_3\right)}{d\alpha_3}\right). \end{aligned} \tag{30}$$

Since both parts of the equation (30) depend on different variables, they are equal to some constant. From left part equation (30) we have equation (20). The equation (20) can to rewrite with account (28) as follows

$$\Phi_2\left(\alpha_2\right)\left(C_4 - \frac{m_1^2}{\sin^2\alpha_2}\right) + \frac{1}{\sin\alpha_2}\frac{d}{d\alpha_2}\left(\sin\alpha_2\frac{d\Phi_2\left(\alpha_2\right)}{d\alpha_2}\right) = 0. \tag{31}$$

The equation (31) is the adjoined Legendre equation. In case $C_4 = l(l+1)$, were l are integers, its solutions are the adjoined Legendre polynomials (Landau & Lifshitz, 1963)

$$P_l^{m_1}\left(\cos\alpha_2\right) = \frac{1}{2^l l!}\sin^{m_1}\alpha_2\frac{d^{m_1+l}}{\left(d\cos\alpha_2\right)^{m_1+l}}\left(\cos^2\alpha_2 - 1\right)^l. \tag{32}$$

So can to get second quantum number l – orbital quantum number.

From right part equation (30) there is

$$C_2\sin^2\alpha_3 + \frac{1}{\Phi_3\left(\alpha_3\right)}\frac{d}{d\alpha_3}\left(\sin^2\alpha_3\frac{d\Phi_3\left(\alpha_3\right)}{d\alpha_3}\right) = \mathrm{C}_4. \tag{33}$$

One rewrite the equation (33) in the form

$$\left(C_2 - l\frac{l+1}{\sin^2\alpha_3}\right)\Phi_3\left(\alpha_3\right) + \frac{1}{\sin^2\alpha_3}\frac{d}{d\alpha_3}\left(\sin^2\alpha_3\frac{d\Phi_3\left(\alpha_3\right)}{d\alpha_3}\right) = 0. \tag{34}$$

Since $\sin^2\alpha_3$ is a positive function varying from 0 to 1, then one introduce the average value of this function τ in the range α_3 from 0 to 2π

$$\tau = \frac{1}{2\pi}\int_0^{2\pi}\sin^2\alpha_3 d\alpha_3 = \frac{1}{2}.$$

Then the equation have form

$$C_2 - 2l\left(l+1\right) = -\frac{1}{\Phi_3\left(\alpha_3\right)}\frac{d^2\Phi_3\left(\alpha_3\right)}{d\alpha_3^{\ 2}}. \tag{35}$$

The decision of this equation is

$$\Phi_3\left(\alpha_3\right) = Be^{\pm i\sqrt{C_2 - 2l(l+1)}\alpha_3}. \tag{36}$$

Since $\Phi_3\left(0\right) = \Phi_3\left(2\pi\right)$, then $B = Be^{\pm i\sqrt{C_2 - 2l(l+1)}2\pi}$, i.e. $e^{\pm i\sqrt{C_2 - 2l(l+1)}2\pi} = 1$.
Consequently, $\cos\left(2\pi\sqrt{C_2 - 2l(l+1)}\right) \pm i\sin\left(2\pi\sqrt{C_2 - 2l(l+1)}\right) = 1$.
It may be only if $\sqrt{C_2 - 2l(l+1)} = 0, \pm 1, \pm 2, \ldots$.
So can to get the third quantum number

$$m_2 = \sqrt{C_2 - 2l(l+1)} = 0, \pm 1, \pm 2, \ldots \tag{37}$$

The constant *B* is defined from condition of normalization

$$\int_0^{2\pi}\Phi_3^*\left(\alpha_3\right)\Phi_3\left(\alpha_3\right)d\alpha_3 = 1 = B^2\int_0^{2\pi}e^{-i\sqrt{C_2 - 2l(l+1)}\alpha_3}e^{i\sqrt{C_2 - 2l(l+1)}\alpha_3}d\alpha_3 = B^2 2\pi.$$

Consequently, $B = \frac{1}{\sqrt{2\pi}}$. Thus, the function $\Phi_3\left(\alpha_3\right)$ it is known

$$\Phi_3\left(\alpha_3\right) = \frac{1}{\sqrt{2\pi}} e^{-im_2\alpha_3}, m_2 = \sqrt{C_2 - 2l(l+1)}. \tag{38}$$

One now turn to equation (24) with respect to the angle α_4 .
The equation (34) can to rewrite as follows

$$\left(C_1 - 2l\frac{l+1}{\sin^2\alpha_4}\right)\Phi_4\left(\alpha_4\right) + \frac{1}{\sin^2\alpha_4}\frac{d}{d\alpha_4}\left(\sin^2\alpha_4\frac{d\Phi_4\left(\alpha_4\right)}{d\alpha_4}\right) = 0. \tag{39}$$

Performing the transformations in equation (39) for the angle α_4 are the same as in equation (34) for the angle α_3, one transform equation (39) to the form

$$C_1 - 4l(l+1) = -\frac{1}{\Phi_4\left(\alpha_4\right)}\frac{d^2\Phi_4\left(\alpha_4\right)}{d\alpha_4^{\ 2}} \tag{40}$$

The equation (40), so equation (35), have decision

$$\Phi_4\left(\alpha_4\right) = \frac{1}{\sqrt{2\pi}} e^{-im_3\alpha_4},$$

were

$$m_3 = \sqrt{C_1 - 4l(l+1)} = 0; \pm 1, \pm 2, \ldots \text{is fourth quantum number.} \tag{41}$$

We now consider the equation for the radius (12). For p = 5 and $C_1 = m_3^2 + 4l(l)$ this equation has the form

$$\frac{1}{Rr^2}\frac{d}{dr}\left(r^4\frac{dR}{dr}\right) + 2r^2\left(\frac{1}{r} + E\right) = m_3^2 + 4l(l+1).$$

Performing differentiation in this equation, can to obtain

$$r^2 \frac{d^2R}{dr^2} + 4r\frac{dR}{dr} + R\left(2r + 2Er^2 - 4l(l+1) - m_3^2\right) = 0. \tag{42}$$

One seek the solution of the equation (42) in the form of a series in powers of *r*

$$R(r) = e^{-\mu r} r^l \sum_j b_j r^j. \tag{43}$$

In (43) $\mu = \sqrt{-2E}, E < 0, b_j$ are numerical coefficients. One substitute (43) into (42)

$$\sum_j b_j r^{l+j}\left((l+j)^2 + 2(l+j) - 4l(l+1) - m_3^2\right) = \sum_j b_j r^{l+j+1}\left(2\mu(l+j+2) - 2\right). \tag{44}$$

From equation (44) follows the recurrence formula for the coefficients

$$b_{j+1} = \frac{2\mu(l+j+2) - 2}{(l+j)^2 + 2(l+j) - 4l(l+1) - m_3^2} b_j. \tag{45}$$

In view of the need to obtain a limited physical meaningful solution, the values of the coefficients, starting from a certain value of *j* must vanish. This is possible if the numerator on the right-hand side of (45) is zero

$$\mu(l+j+2) - 1 = 0 \tag{46}$$

From equation (46) follows that

$$\mu = \sqrt{-E} = \frac{1}{l+j+2} \tag{47}$$

One introduces the notation $n = l + j + 1$ is principal quantum number. From (47) equation can to obtain

$$E = -\frac{1}{2(n+1)^2} \tag{48}$$

The radial function is determined by the equality (43) of setting different values *j*. For example, with *j* = 0

$$R(r) = \mathrm{e}^{-\mu r} r^l b_0, b_0 = \frac{1}{\int_0^\infty e^{-\mu r} r^l dr}$$

With *j* = *1*

$$R(r) = e^{-\mu r} r^l b_0 \left(1 - r\frac{2\mu(l+2) - 2}{3l^2 + 2l + m_3^2}\right),$$

$$b_0 = \frac{1}{\int_0^\infty e^{-\mu r} r^l \left(1 + \frac{2 - 2\mu(l+2)}{3l^2 + 2l + \mathrm{m}_3^2}\right) dr}$$

Thus, one can obtain various approximations for the function $R(r)$.

CONCLUSION

Thus, it is proved that the Schrödinger equation has a solution in a space of higher dimension. Consequently, the discrete nature of matter exists in a space of higher dimension. The incorrectness of authors in previous studies of the solution of the Schrodinger equation in space of higher dimension was that the potential energy of electron was written in form different at law of Coulomb. The study showed that the motion of an electron around a nucleus in a space of higher dimension leads to the appearance of quantum numbers with a total number equal to the dimension of space. On a level with the principal quantum number, characterizing the average distance of the electron from the nucleus, and the orbital quantum number, characterizing the shape of the electron orbitals, a series of magnetic quantum numbers appear in a space of higher dimension. Its correspondent the possible values of the projections of the angular momentum of an electron on the coordinate axes of space. The number of these magnetic quantum numbers is equal to the dimension of the space minus two. Taking into account the principal quartic number and orbital quantum number, the total number of quantum numbers characterizing the motion of an electron around the nucleus is equal to the dimension of space. Thus, the total number of quantum numbers of an electron with allowance for the spin quantum number characterizing the internal rotation of an electron is equal to the dimension of space plus one. In this case, the electron energy in the first approximation, as well as in the model of the Bohr atom, is inversely proportional to the square of the principal quantum number (equation (48)). For example, in three - dimensional space the atom has besides a spin quantum number yet of three more quantum numbers. The main quantum number n, which determines the average distance of the electron from the nucleus. Orbital quantum number l, characterizing the momentum of an electron. The magnetic quantum number m_l, characterizing the possible value of the projection of the angular momentum of the electron on the axis z in a magnetic field. The main quantum number can take the values 1, 2, 3, ... The orbital quantum number can take the values 0, 1, 2, ...

Magnetic quantum numbers can take the values 0, ±1, ±2, ±3,...

The values of the orbital quantum number l correspond to form of electron orbital. Taking into account the large dimensionality of the neighborhood of the nucleus in the atom, one should expect an increase in the number of quantum cells in orbitals *p, d, f* while maintaining their shape. For example, not 6 but 8 electrons (for each quantum cell with a corresponding pair of electrons per axis) can be on the orbital *p*. When analyzing chemical elements and molecules formed by them, it was shown (Zhizhin, 2017) that many elements exhibit a valence in chemical reactions much higher than the valence determined by their place in the periodic table (group number). To explain this, can to resort to assumptions about various possible mechanisms of electron interaction: atomic and molecular hybridization, repulsion of divided and unpaired electron pairs, and the involvement of electrons inside the electron shells in the chemical interactions. In addition, physical experiments at high pressures reveal a sharp change in the structure of compounds that are also not explainable from the point of view of standard valence.

On the other hand, the analysis of real forms of molecules shows that their dimensionality, determined by the shape of the corresponding convex body, whose vertices are atoms, is often higher than three.

Undoubtedly, the possible loss of the number of quantum cells in orbitals *p, d, f* should not lead to a change in the number of electrons in the atomic structure in neutral atoms. The number of electrons under these conditions is of course equal to the charge of the nucleus. But the possible increase in the number of vacant quantum cells contributes to an increase in the chemical activity of atoms, allows the creation of complex chemical compounds. This can to see, for example, in biology.

REFERENCES

Babenko, Yu. I. (2015). Antispace. In I.N. Toganov, Yu.I. Babenko. Anti-time and antispace. S.-Petersburg: TIN, 187-238.

Büchel, W. (1963). Warum hat ger Raume drei Dimensionen? *Physilalische Blatter*, *19*(12), 547–549. doi:10.1002/phbl.19630191204

Freeman, I. M. (1969). Why is Space Three-Dimensional? *American Journal of Physics*, *37*, 1222–1224. doi:10.1119/1.1975283

Gurevich, L., & Mostepanenko, V. (1971). On the Existence of Atoms in *n* - Dimensional Space. *Physics Letters. [Part A]*, *35*(3), 201–202. doi:10.1016/0375-9601(71)90148-4

Hobson, E. W. (1931). *The theory of spherical and allipsoidal harmonics*. Cambridge: At the University Press.

Landau, L. D., & Lifshitz, E. M. (1963). Quantum mechanics. Nonrelativistic theory. Moscow: Gos. Ed. Phys.-Math. Lit.

Poincare, A. (1895). Analysis Situs. *Journal de Ecole Polytechnique*, *1*, 1–121.

Schrodinger. (1928). *Abhandlungen zur Wellenmechanik*. Leipzig.

Zhizhin, G. V. (2017). *Chemical Compound Structures and the Higher Dimension of Molecules: Emerging Research and Opportunities*. Hershey, PA: IGI Global.

Chapter 9
The Partition of *n* - Dimensional Space of Polytopic Prismahedrons

ABSTRACT

It is proved that polytopic prismahedrons have the necessary properties for partitioning the n - dimensional spaces of a face into a face, that is, they satisfy the conditions for solving the eighteenth Hilbert problem of the construction of n - dimensional spaces from congruent figures. General principles and an analytical method for constructing n -dimensional spaces with the help of polytopic prismahedrons are developed. On the example of specific types of the polytopic prismahedrons (tetrahedral prism, triangular prismahedron), the possibility of such constructions is analytically proved. It was found that neighboring polytopic prismahedrons in these constructions can have common geometric elements of any dimension less than n or do not have common elements.

INTRODUCTION

The problem of completing space by polyhedrons is one of the fundamental problems of mathematics, which has long attracted the attention of scientists. In 1900, D. Gilbert formulated 23 mathematical problems that require solution (Gilbert, 1901). One of these problems (eighteenth) was devoted to this question. It was formulated as follows: "Construction of space from

DOI: 10.4018/978-1-5225-6968-8.ch009

congruent polyhedrons." This problem is especially complicated in the case of n-dimensional spaces (Delone, 1969), and up to the end it has not been solved to this day under these conditions. In 1961, Delone (Delone, 1961) proved that if we require that polytopes in n - dimensional space be adjacent along entire (n - 1) faces, then for any n there are only finitely many topologically different partitions of space into polytopes. These partitions are called normal. Delone and Sandakova (Delone & Sandakova, 1961) obtained a finite algorithm, in principle (according to the authors), which allows one to find all such partitions for a given n. In this case, the polytopes themselves (stereohedrons by the Delone terminus) of these partitions for a fixed n can be only of a finite number of topologically different types. However, unfortunately, the authors did not obtain concrete examples of normal partitions of higher - dimensional space into polytopes and the specific types of corresponding polytopes. The question of the existence of fundamental domains in n - dimensional space is quite nontrivial. The main definition of the fundamental domain it was formulated by Delone (Delone, 1969). The fundamental domain of a group of motions is a set of points of space such that, firstly, all its points are not equivalent to each other with respect to the group of motions, and secondly, any point of space is equivalent to some point of this region relative to the group of motions. What kind of these fundamental areas for a space with a dimension greater than three remains to this day is unknown. Here we must also keep in mind the existence of polytopes, whose congruential copies can fill a gapless space, but each of them is not a fundamental area (Reingard, 1928).

The partitions of higher dimensional space first it were given by Zhizhin (Zhizhin, 2017). He showed that polytopes of partitions of higher - dimensional space are the polytopic prismahedrons. Moreover, the normality condition in these partitions is not obligatory, i.e. in the n - dimensional space there are polytopes adjacent along entire (n - 1) faces, but there are polytopes adjacent along the faces of another dimension.

One begin our consideration of the problem of partitioning a space of higher dimension from the reduction of concrete examples of these partitions.

EXAMPLES OF PARTITIONS OF HIGHER DIMENSIONAL SPACE

Partition of Space by the Polytopic Prism

One of the simplest types of the polytopic prismahedrons is the tetrahedral prism, i.e. a product tetrahedron by the segment (Chapter 4). Its structural formula has form (8) of Chapter 4

$$P_4^3(4F_3^2)\times P_2^1 = P_8^4[4P_6^3(3F_4^2,2F_3^2),2F_4^3(4F_3^2)]. \tag{1}$$

Here, the subscript in the polytope P and his face F indicates the number of vertices, and the superscript indicates the dimension of the corresponding polytope or faces. The right side of (1) describing the structural formula of the product, the facet indicated by the symbol of the polytope to specify which polytopes of dimension n - 1 is composed work polytopes. Thus, P_4^3 - the tetrahedron, P_2^1 - the segment, P_6^3 - the triangular prism, P_3^2 - the triangle, P_4^2 - the quadrilateral, P_8^4 - the tetrahedral prism. The dimension of the tetrahedral prism is equal to 4, it has 8 vertices, 16 edges, 14 faces two - dimensional, 6 three - dimensional faces (2 tetrahedrons, 4 triangular prisms). Image tetrahedral prism is shown in Figure 1 in Chapter 4.

Can to introduce one of the vertices of the tetrahedral prism origin of the four - dimensional space (x, y, z, t). Orient the coordinates, such as indicated in Figure 1.

Assume that the length of each edge is equal to 1. Then, each node tetrahedral prism can be associated with a set of integers (Figure 1). Translating tetrahedral prism along the coordinates x, y, z, t, we obtain the lattice vertices. Let A_0 tetrahedral prism with the values of vertex coordinates in Figure 1. Then

$$A_0 = [(0,0,0,0),(1,0,0,0),(0,1,0,0),(0,0,1,0),(0,0,0,1),(1,1,0,0),(0,1,1,0),(0,1,0,1)], \tag{2}$$

$$A_1 = A_0(x+1) = [(1,0,0,0),(2,0,0,0),(1,1,0,0),(1,0,1,0),(1,0,0,1),(2,1,0,0),(1,1,1,0),(1,1,0,1)],$$

Figure 1. The tetrahedral prism

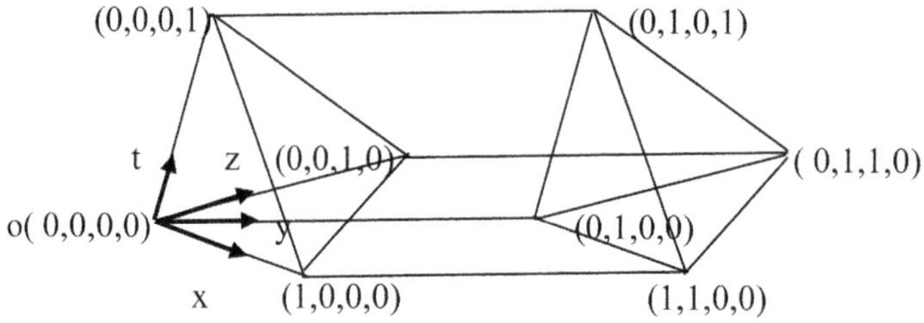

$$A_2 = A_0(z+1) = [(0,0,1,0),(1,0,1,0),(0,1,1,0),(0,0,2,0),(0,0,1,1),(1,1,1,0),(0,1,2,0),(0,1,1,1)],$$

$$A_3 = A_0(x+1, z+1) = A_1(z+1) = \\ [(1,0,1,0),(2,0,1,0),(1,1,1,0),(1,0,2,0),(1,0,1,1,),(2,1,1,0),(1,1,2,0),(1,1,1,1)],$$

$$A_4 = A_0(y+1, z+1) = A_2(y+1) = \\ [(0,1,1,0),(1,1,1,0),(0,2,1,0),(0,1,2,0),(0,1,1,1),(1,2,1,0),(0,2,2,0),(0,2,1,1)],$$

$$A_5 = A_0(x+1, y+1, z+1) = A_3(y+1) = \\ [(1,1,1,0),(2,1,1,0),(1,2,1,0),(1,1,2,0),(1,1,1,1),(2,2,1,0),(1,2,2,0),(1,2,1,1)],$$

$$A_6 = A_0(y+1) = [(0,1,0,0),(1,1,0,0),(0,2,0,0),(0,1,1,0),(0,1,0,1),(1,2,0,0),(0,2,1,0),(0,2,0,1)],$$

$$A_7 = A_0(y+1, x+1) = A_1(y+1) = \\ [(1,1,0,0),(2,1,0,0),(1,2,0,0),(1,1,1,0),(1,1,0,1),(2,2,0,0),(1,2,1,0),(1,2,0,1)],$$

$$A_8 = A_0(t+1) = \\ [(0,0,0,1),(1,0,0,1),(0,1,0,1),(0,0,1,1),(0,0,0,2),(1,1,0,1),(0,1,1,1),(0,1,0,2)],$$

$$A_9 = A_0(y+1, t+1) = A_6(t+1) = \\ [(0,1,0,1),(1,1,0,1),(0,2,0,1),(0,1,1,1),(0,1,0,2),(1,2,0,1),(0,2,1,1),(0,2,0,2)],$$

$$A_{10} = A_0(x+1, y+1, t+1) = A_7(t+1) =$$
$$[(1,1,0,1),(2,1,0,1),(1,2,0,1),(1,1,1,1),(1,1,0,2),(2,2,0,1),(1,2,1,1),(1,2,0,2)],$$

$$A_{11} = A_0(x+1, t+1) = A_1(t+1) =$$
$$[(1,0,0,1),(2,0,0,1),(1,1,0,1),(1,0,1,1),(1,0,0,2),(2,1,0,1),(1,1,1,1),(1,1,0,2)],$$

$$A_{12} = A_0(y+1, z+1) = A_2(t+1) =$$
$$[(0,0,1,1),(1,0,1,1),(0,1,1,1),(0,0,2,1),(0,0,1,2),(1,1,1,1),(0,1,2,1),(0,1,1,2)],$$

$$A_{13} = A_0(z+1, y+1, t+1) = A_4(t+1) =$$
$$[(0,1,1,1),(1,1,1,1),(0,2,1,1),(0,1,2,1),(0,2,2,1),(0,1,1,2),(1,2,1,1),(0,2,1,2)],$$

$$A_{14} = A_0(x+1, y+1, z+1, t+1) = A_5(t+1) =$$
$$[(1,1,1,1),(2,1,1,1),(1,2,1,1),(1,1,2,1),(1,1,1,2),(2,2,1,1),(1,2,2,1),(1,2,1,2)],$$

$$A_{15} = A_0(x+1, t+1, z+1) = A_3(t+1) =$$
$$[(1,0,1,1),(2,0,1,1),(1,1,1,1),(1,0,2,1),(1,0,1,2),(2,1,1,1),(1,1,2,1),(1,1,1,2)].$$

Representing the tetrahedral prisms $A_0 \div A_{15}$ dots in three - dimensional space, can to get the hypercube. Moreover, the edges of the hypercube correspond to possible changes in the values of one of the coordinates of the vertices of the tetrahedral prism unit (Figure 2).

In addition, each edge of the hypercube in Figure 2 can be considered as an element of the overall two tetrahedral prisms, connected by an edge. Using the coordinate expression tetrahedral prisms (2) can be analytically determined. In Table 1 geometry elements are common to each pair of the tetrahedral prisms connected by an edge in Figure 2 are listed.

Diagonals flat faces in the hypercube correspond to a simultaneous change in the unit values of the two coordinates of the vertices of the tetrahedral prisms (Table 2). Diagonals 8 cubes in the hypercube in Figure 2 correspond to a simultaneous change in the unit of some three coordinates. Common elements of the tetrahedral prisms with such changes vertex coordinates is either a vertex or the empty set. When you change the same unit coordinates all four common elements in the tetrahedral prisms not.

Figure 2. The hypercube from 16 tetrahedral prisms

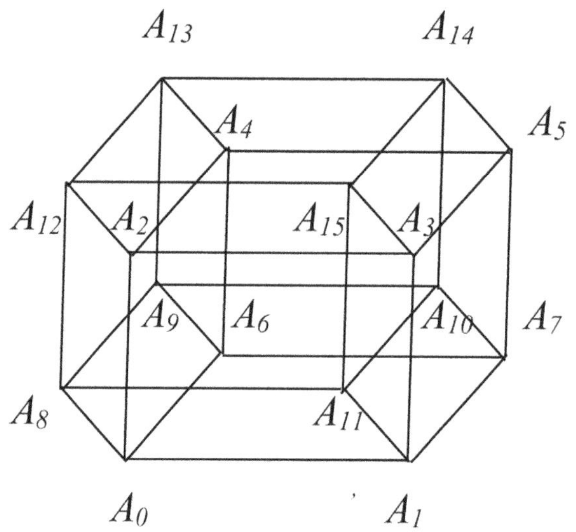

Between A_i type tetrahedral prisms arranged tetrahedral prism type B_i (see Figure 3) having a closest to them tetrahedral prisms A_i general quadrangular two - dimensional face.

The tetrahedral prisms A_i and B_i have common edges, which are parties to the general quadrilateral faces. The tetrahedral prisms A_i and B_i are connected symmetry transformation - turning the 180^0 around a common edge of the tetrahedron. Let B_0 tetrahedral prism type B_i, adjacent tetrahedral prism A_0, contains the x - axis, will translate the prism B_0 on edge length for all the coordinates x, y, z, t four - dimensional space. Then we obtain a lattice of the tetrahedral prisms B_i.

Figure 3. The partition space by the tetrahedral prisms

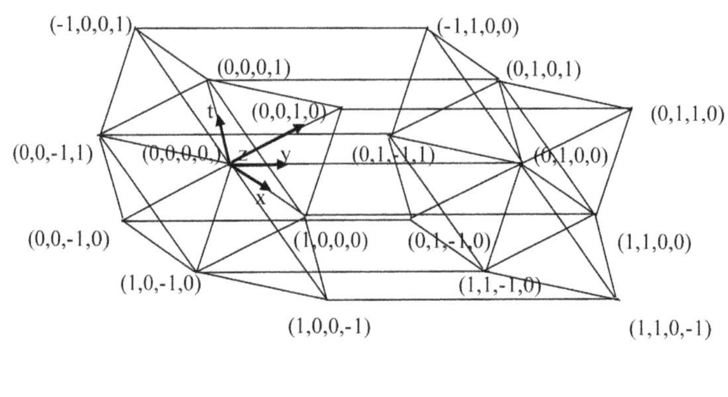

$$B_0 = \begin{bmatrix} (0,0,0,0),(0,1,0,0),(1,1,0,0),(1,0,0,0), \\ (1,0,-1,0),(1,0,0,-1),(1,1,0,-1),(1,1,-1,0) \end{bmatrix}, \tag{3}$$

$$B_1 = B_0(x+1) = \begin{bmatrix} (1,0,0,0),(1,1,0,0),(2,1,0,0),(2,0,0,0), \\ (2,0,-1,0),(2,0,0,-1),(2,1,0,-1),(2,1,-1,0) \end{bmatrix},$$

$$B_2 = B_0(z+1) = \begin{bmatrix} (0,0,1,0),(0,1,1,0),(1,1,1,0),(1,0,1,0), \\ (1,0,0,0),(1,0,1,-1),(1,1,1,-1),(1,1,0,0) \end{bmatrix},$$

$$B_3 = B_0(x+1,z+1) = B_1(z+1) = \begin{bmatrix} (1,0,1,0),(1,1,1,0),(2,1,1,0),(2,0,1,0), \\ (2,0,0,0,),(2,0,1,-1),(2,1,1,-1),(2,1,0,0) \end{bmatrix},$$

$$B_4 = B_0(y+1,z+1) = B_2(y+1) = \begin{bmatrix} (0,1,1,0),(0,2,1,0),(1,2,1,0),(1,1,1,0), \\ (1,1,0,0),(1,1,1,-1),(1,2,1,-1),(1,2,0,0) \end{bmatrix},$$

$$B_5 = B_0(\mathrm{x}+1,y+1,z+1) = B_3(y+1) = \begin{bmatrix} (1,1,1,0),(1,2,1,0),(2,2.1.0),(2,1,1,0), \\ (2,1,0,0),(2,1,1,-1),(2,2,1,-1),(2,2,0,0) \end{bmatrix},$$

$$B_6 = B_0(y+1) = \begin{bmatrix} (0,1,0,0),(0,2,0,0),(1,2,0,0),(1,1,0,0), \\ (1,1,-1,0),(1,1,0,-1),(1,2,0,-1),(1,2,-1,0) \end{bmatrix},$$

$$B_7 = B_0(y+1,\mathrm{x}+1) = B_1(y+1) =$$
$$[(1,1,0,0),(1,2,0,0),(2,2,0,0),(2,1,0,0),(2,1,-1,0),(2,1,0,-1),(2,2,0,-1),(2,2,-1,0)],$$

$$B_8 = B_0(\mathrm{t}+1) =$$
$$[(0,0,0,1),(0,1,0,1),(1,1,0,1),(1,0,0,1),(1,0,-1,1),(1,0,0,1),(1,1,0,0),(1,1,-1,1)],$$

$$B_9 = B_0(y+1, \mathrm{t}+1) = B_6(\mathrm{t}+1) =$$
$$[(0,1,0,1),(0,2,0,1),(1,2,0,1),(1,1,0,1),(1,1,-1,1),(1,1,0,0),(1,2,0,0),(1,2,-1,1)],$$

$$B_{10} = B_0(\mathrm{x}+1, y+1, \mathrm{t}+1) = B_7(\mathrm{t}+1) =$$
$$[(1,1,0,1),(1,2,0,1),(2,2,0,1),(2,1,0,1),(2,1,-1,1),(2,1,0,0),(2,2,0,0),(2,2,-1,1)],$$

$$B_{11} = B_0(\mathrm{x}+1, \mathrm{t}+1) = B_1(\mathrm{t}+1) =$$
$$[(1,0,0,1),(1,1,0,1),(2,1,0,1),(2,0,0,1),(2,0,-1,1),(2,0,0,0),(2,1,0,0),(2,1,-1,1)],$$

$$B_{12} = B_0(\mathrm{y}+1, \mathrm{z}+1) = B_2(\mathrm{t}+1) =$$
$$[(0,0,1,1),(0,1,1,1),(1,1,1,1),(1,0,1,1),(1,0,0,1),(1,0,1,0),(1,1,1,0),(1,1,0,1)],$$

$$B_{13} = B_0(\mathrm{z}+1, \mathrm{y}+1, t+1) = B_4(\mathrm{t}+1) =$$
$$[(0,1,1,1),(0,2,1,1),(1,2,1,1),(1,1,1,1),(1,1,0,1),(1,1,1,0),(1,2,1,0),(1,2,0,1)],$$

$$B_{14} = B_0(\mathrm{x}+1, \mathrm{y}+1, z+1, t+1) = B_5(\mathrm{t}+1) =$$
$$[(1,1,1,1),(1,2,1,1),(2,2,1,1),(2,1,1,1),(2,1,0,1),(2,1,1,0),(2,2,1,0),(2,2,0,1)],$$

$$B_{15} = B_0(\mathrm{x}+1, \mathrm{t}+1, z+1) = B_3(\mathrm{t}+1) =$$
$$[(1,0,1,1),(1,1,1,1),(2,1,1,1),(2,0,1,1),(2,0,0,1),(2,0,1,0),(2,1,1,0),(2,1,0,1)].$$

It is presented tetrahedral prisms $B_0 \div B_{15}$ points in the 4 - dimensional space can to get up hypercube in Figure 2 with change notation A_i on B_i. From the construction it is follows that tetrahedral prism B_i can have common elements such as tetrahedrons, edges and vertices. This is easily seen by analytical expressions (3). In addition, it follows from the construction (Figure 3) that the tetrahedral prism A_i and B_i can have common elements only tetrahedrons, edges and vertices. Triangular prisms, which are present in tetrahedral prisms of both types can not be common for these tetrahedral prisms.

They only deal with each other on a flat quadrilateral. This also easily seen from the formulas (2) and (3). This situation is significantly different from the conditions of normality, to accept in general theory of stereohedrons. The provided construction proves that the tetrahedral prism is a fundamental area in the 4 - dimensional space, as it completely fills the space in the translation of the prism in all the coordinates of space and turning it into 180^0.

Since all the vertices of the tetrahedral prisms are equal and filling the space with their translation is uniform, then the partition created by them is the right one space.

As follows from the division of space by the tetrahedral prisms (Figure 3), the geometric shape of the tetrahedral prism *B* (Figure 4) differs from the geometric shape of the tetrahedral prism *A*.

Table 1. Common elements of the tetrahedral prisms

Edge of the Hypercube	Common Element of the Tetrahedral Prisms	Edge of the Hypercube	Common Element of the Tetrahedral Prisms
A_0A_1	edge (1,0,0,0)(1,1,0,0)	A_5A_{14}	edge (1,1,1,1)(1,2,1,1)
A_0A_2	edge (0,1,1,0)(0,0,1,0)	A_6A_7	vertex (1,1,1,0)
A_0A_6	tetrahedron (0,1,0,0)(1,1,0,0)(0,1,0,1)(0,1,0,1)	A_6A_9	edge (0,1,0,1)(0,2,0,1)
A_0A_8	edge (0,1,0,1)(0,0,0.1)	A_7A_{10}	vertex (1,1,0,1)
A_1A_3	edge (1,0,1,0)(1,1,1,0)	A_8A_{11}	edge (1,0,0,1)(1,1,0,1)
A_7A_1	tetrahedron (1,1,0,0,)(2,1,0,0)(1,1,1,0)(1,1,0,1)	A_8A_9	tetrahedron (0,1,0,1)(1,1,0,1)(0,1,1,1)(0,1,0,2)
A_1A_{11}	edge (1,0,0,1)(1,1,0,1)	A_8A_{12}	edge (0,0,1,1)(0,1,1,1)
A_2A_3	edge (1,0,1,0)(1,1,0,0)	A_9A_{10}	edge (1,1,0,1)(1,2,0,1)
A_2A_4	tetrahedron (0,1,1,0)(1,1,1,0)(0,1,1,1)(0,1,2,0)	A_9A_{13}	edge (1,2,1,1)(0,1,1,1)
A_2A_{12}	edge (0,1,1,0)(0,0,1,1)	$A_{11}A_{10}$	edge (1,2,0,1)(1,1,0,1)
A_3A_5	tetrahedron (1,1,1,0)(2,1,1,0)(1,1,2,0)(0,1,1,1)	$A_{14}A_{10}$	edge (1,1,1,1)(1,2,1,1)
A_3A_{15}	edge (1,0,1,1)(1,1,1,1)	$A_{11}A_{15}$	edge (1,0,1,1)(1,1,1,1)
A_4A_5	edge (1,1,1,0)(1,2,1,0)	$A_{13}A_{12}$	edge (0,1,1,1)(1,1,1,1)
A_4A_6	edge (0,1,1,0)(0,2,1,0)	$A_{12}A_{15}$	edge (1,0,1,1)(1,1,1,1)
A_4A_{13}	vertex (0,1,1,1)	$A_{13}A_{14}$	tetrahedron (1,1,1,1)(1,1,2,1)(1,2,1,1)(1,2,2,1)
A_7A_5	edge (1,1,1,0)(1,2,1,0)	$A_{14}A_{15}$	tetrahedron (1,1,1,1)(2,1,1,1)(1,1,2,1)(1,1,1,1)

Table 2. Common elements of the tetrahedral prisms

Diagonal Flat Face of the Hypercube	Common Element of the Tetrahedral Prisms	Diagonal Flat Face of the Hypercube	Common Element of the Tetrahedral Prism
A_8A_{13}, A_8A_{13}	vertex (0,1,1,1)	A_8A_{10}, A_9A_{11}	vertex (1,1,1,1)
A_6A_2, A_0A_4	vertex (0,1,1,0)	$A_{14}A_7$	∅
A_8A_2	edge (0,1,1,1)(0,0,1,1)	A_5A_{10}	edge (1,1,1,1)(1,2,1,1)
A_0A_{12}	∅	$A_{13}A_{14}$, A_5A_{15}	vertex (1,1,1,1)
A_4A_9	vertex (0,1,1,1)(0,0,1,1)	A_3A_{11}	vertex (1,0,1,1)(1,1,1,1)
A_6A_{13}	∅	A_1A_{15}	∅
A_5A_{13}	edge (1,1,1,1)(1,2,1,1)	A_7A_{11}, A_1A_{10}	vertex (1,1,0,1)
A_4A_{14}	∅	A_5A_1, A_7A_3	vertex (1,1,1,0)
A_5A_{10}	edge (1,2,1,1)(1,1,1,1)	A_3A_{12}	edge (1,0,1,1)(1,1,1,1)
A_7A_{14}	∅	A_2A_{15}	∅
A_7A_9	edge (1,1,0,1)(1,2,0,1)	$A_{11}A_0$	∅
A_6A_{10}	∅	A_1A_8	edge (1,0,0,1)(1,1,0,1)
A_9A_{14}	∅	A_4A_3, A_5A_2	edge (1,1,1,0)
$A_{13}A_{10}$	edge (1,2,1,1)(1,1,1,1)	A_1A_6, A_0A_7	vertex (1,1,0,0)
A_5A_6	∅	A_6A_8, A_0A_9	vertex (0,1,0,1)
A_4A_7	edge (1,1,1,0)(1,2,1,0)	$A_{11}A_{14}$, $A_{10}A_{15}$	vertex (1,1,1,1)
$A_{11}A_{12}$	edge (1,1,1,1)(1,0,1,1)	A_8A_{15}	∅
$A_{14}A_{12}$	vertex (1,1,1,1)	$A_{13}A_{15}$	edge (1,1,2,1)(1,1,1,1)

Figure 4. The tetrahedral prism B

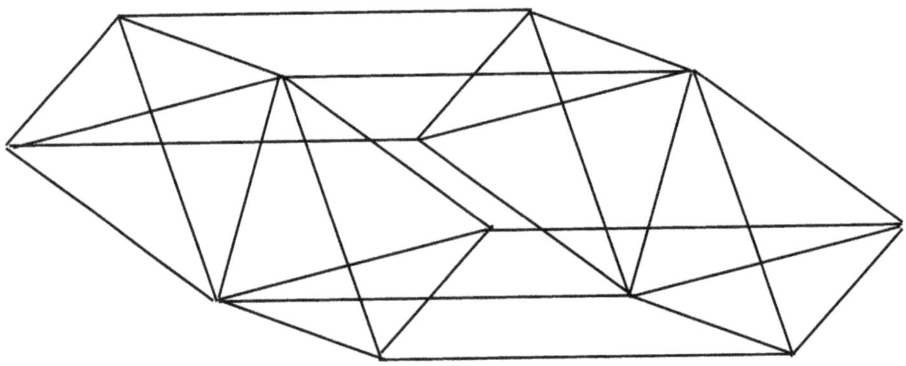

These differences are related to the different angle of the one - dimensional segment, as a factor of the tetrahedral prism, to the tetrahedron in these tetrahedral prisms. Moreover, the angles between tetrahedrons and one - dimensional segment are coordinated so that tetrahedral prisms A and B can adjoin to each other along a common two - dimensional face having the form of a parallelogram. The polytope depicted in the figure consists of four tetrahedral prisms: two tetrahedral prisms of type A and two tetrahedral prisms of type B. The tetrahedral prisms A and B have common two - dimensional quadrangular faces, and each of the pairs of tetrahedral prisms A and B have the common one - dimensional segment. This polytope, as is easy to see, is the fundamental domain of the partition of a four - dimensional space by tetrahedral prisms, since the translation of this region covers all four - dimensional space. Moreover, translation in the direction of a one - dimensional segment, which is a factor in the product of tetrahedrons per segment, leads to overlapping of regions with four tetrahedrons, and translation in directions perpendicular to this segment leads to overlapping of regions along two - dimensional faces. Moreover, each point of this region does not have an equivalent in the region with respect to the enumerated group of motions. Thus, for the first time the concrete fundamental region of four - dimensional space constructed has a complex form in the case of tetrahedral prisms and is a complex of tetrahedral prisms. Each individual tetrahedral prism is not a fundamental domain.

The fundamental area of the partition of a four - dimensional space by tetrahedral prisms can be conditionally represented as a set of tetrahedral prisms $\begin{matrix} BA \\ AB \end{matrix}$ in which the diagonal elements are joined to each other along the edge, and the elements along the vertical and horizontally join each other along two - dimensional quadrangular surfaces connected along this edge. The fundamental region with adjacent fundamental regions is bordered by two - dimensional quadrangular surfaces and by the totality of four tetrahedrons.

Partition of Space of the Triangular Prismahedron

The triangular prismahedron is the product of a triangle by a triangle. The dimension of this polytope equal 4, it has 9 vertices and 6 triangle prisms (see Figure 5 in Chapter 4). The structural formula of the triangular prismahedron defined of equation (16) in Chapter 4.

One direct the coordinates of the four - dimensional space along the edges of the triangular prismahedron (see the Figure 5).

Figure 5. The triangular prismahedron

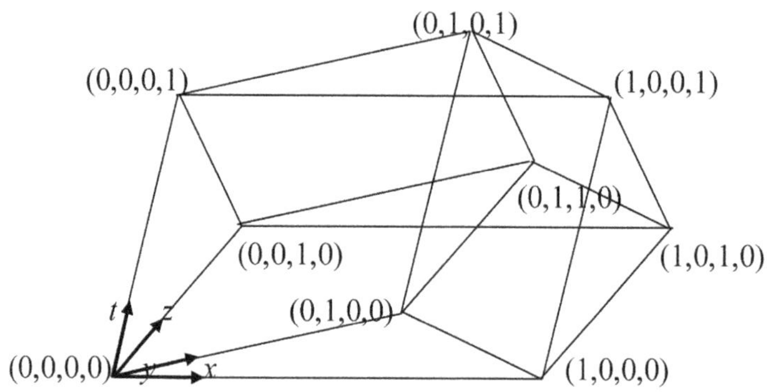

Assume that the length of each edge is equal to 1. Then, each node triangular prismahedron can be associated with a set of integers (Figure 5). Translating the triangular prismahedron along the coordinates *x, y, z, t*, can to obtain the lattice vertices. Let E_0 triangular prismahedron with the values of vertex coordinates in Figure 5. Then

$$E_0 = \begin{bmatrix} (0,0,0,0),(1,0,0,0),(0,1,0,0),(0,0,1,0), \\ (0,0,0,1),(1,0,1,0),(0,1,1,0),(1,0,0,1),(0,1,0,1) \end{bmatrix}, \tag{4}$$

$$E_1 = E_0(x+1) = \begin{bmatrix} (1,0,0,0),(2,0,0,0),(1,1,0,0),(1,0,1,0),(1,0,0,1), \\ (2,0,1,0),(1,1,1,0),(2,0,0,1),(1,1,0,1) \end{bmatrix},$$

$$E_2 = E_0(z+1) = \begin{bmatrix} (0,0,1,0),(1,0,1,0),(0,1,1,0),(0,0,2,0), \\ (0,0,1,1),(1,0,2,0),(0,1,2,0),(1,0,1,1),(0,1,1,1) \end{bmatrix},$$

$$E_3 = E_0(x+1,z+1) = E_1(z+1) = \\ [(1,0,1,0),(2,0,1,0),(1,1,1,0),(1,0,2,0),(1,0,1,1,),(2,0,2,0),(1,1,2,0),(2,0,1,1),(1,1,1,1)],$$

$$E_4 = E_0(y+1,z+1) = E_2(y+1) = \\ [(0,1,1,0),(1,1,1,0),(0,2,1,0),(0,1,2,0),(0,1,1,1),(1,1,2,0),(0,2,2,0),(1,1,1,1),(0,2,1,1)],$$

$$E_5 = E_0(x+1, y+1, z+1) = E_3(y+1) =$$
$$[(1,1,1,0),(2,1,1,0),(1,2,1,0),(1,1,2,0),(1,1,1,1),(2,1,2,0),(1,2,2,0),(1,2,1,1),(1,2,1,1)],$$

$$E_6 = E_0(y+1) = [(0,1,0,0),(1,1,0,0),(0,2,0,0),(0,1,1,0),(0,1,0,1),(1,1,1,0),(0,2,1,0),(1,1,0,1),$$
$$(0,2,0,1)],$$

$$E_7 = E_0(y+1, \mathrm{x}+1) = E_1(y+1) =$$
$$[(1,1,0,0),(2,1,0,0),(1,2,0,0),(1,1,1,0),(1,1,0,1),(2,1,1,0),(1,2,1,0),(2,1,0,1),(1,2,0,1)],$$

$$E_8 = E_0(t+1) =$$
$$[(0,0,0,1),(1,0,0,1),(0,1,0,1),(0,0,1,1),(0,0,0,2),(1,0,1,1),(0,1,1,1),(1,0,0,2),(0,1,0,2)],$$

$$E_9 = E_0(y+1, t+1) = E_6(t+1) =$$
$$[(0,1,0,1),(1,1,0,1),(0,2,0,1),(0,1,1,1),(0,1,0,2),(1,1,1,1),(0,2,1,1),(1,1,0,2),(0,2,0,2)],$$

$$E_{10} = E_0(x+1, y+1, t+1) = E_7(t+1) =$$
$$[(1,1,0,1),(2,1,0,1),(1,2,0,1),(1,1,1,1),(1,1,0,2),(2,1,1,1),(1,2,1,1),(2,1,0,2),(1,2,0,2)],$$

$$E_{11} = E_0(x+1, t+1) = E_1(t+1) =$$
$$[(1,0,0,1),(2,0,0,1),(1,1,0,1),(1,0,1,1),(1,0,0,2),(2,0,1,1),(1,1,1,1),(2,0,0,2),(1,1,0,2)],$$

$$E_{12} = E_0(y+1, z+1) = E_2(t+1) =$$
$$[(0,0,1,1),(1,0,1,1),(0,1,1,1),(0,0,2,1),(0,0,1,2),(1,0,2,1),(0,1,2,1),(1,0,1,2),(0,1,1,2)],$$

$$E_{13} = E_0(z+1, y+1, t+1) = E_4(t+1) =$$
$$[(0,1,1,1),(1,1,1,1),(0,2,1,1),(0,1,2,1),(0,1,1,2),(0,1,1,2),(1,1,2,1),(1,1,1,2),(0,2,1,2)],$$

$$E_{14} = E_0(x+1, y+1, z+1, t+1) = E_5(t+1) =$$
$$[(1,1,1,1),(2,1,1,1),(1,2,1,1),(1,1,2,1),(1,1,1,2),(2,1,2,1),(1,2,2,1),(2,1,1,2),(1,2,1,2)],$$

$$E_{15} = E_0(x+1, t+1, z+1) = E_3(t+1) =$$
$$[(1,0,1,1),(2,0,1,1),(1,1,1,1),(1,0,2,1),(1,0,1,2),(2,0,2,1),(1,1,2,1),(2,0,1,2),(1,1,1,2)].$$

Representing the triangular prismahedrons $E_0 \div E_{15}$ dots in three-dimensional space, can to get the hypercube. Moreover, the edges of the hypercube correspond to possible changes in the values of one of the coordinates of the vertices of the triangular prismahedron unit (Figure 6).

In addition, each edge of the hypercube in Figure 6 can be considered as an element of the overall two triangular prismahedrons, connected by an edge. Using the coordinate expression triangular prismahedrons (4) can be analytically determined. In Table 3 geometry elements are common to each pair of triangular prismahedrons connected by an edge in Figure 6 are listed. From Table 3 it follows that triangular prismahedrons of type *E* when translating along the coordinates of a four-dimensional space form a network, adjoining each other along entire triangles. The direct construction of this network shows (see Figure 7) that between these prismhedrons formed

Figure 6. The hypercube from 16 triangular prismahedrons

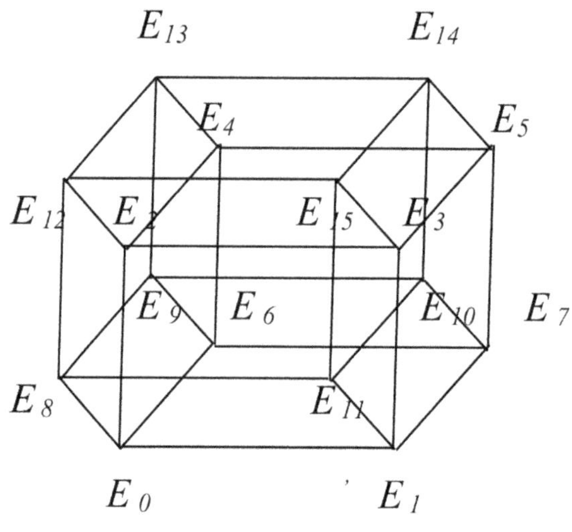

closed areas of three more types, which are also triangular prismahedrons *E*, but differing in the different arrangement of the triangles, as the factors of the product.

The types of these triangular prismahedrons (denoted by *L, C, D*) are presented separately in the Figure 8, Figure 9, Figure 10 and are indicated by a different color.

The triangular prismahedrons *L, C, D*, as well as triangular prismahedrons *E* form networks in which they adjoin each other along entire triangles. This follows from indicated the triangular prismahedrons *L, C, D* in coordinates of vertices of network and tables 4, 5, 6 of common elements in each of network.

The indicated of the triangular prismahedrons type *L*

$$L_0 = [(0,0,1,0),(0,0,0,1),(0,0,1,1),(1,0,1,0),(0,1,1,0),(0,1,0,1),(1,0,0,1),(0,1,1,1),(1,0,1,1),$$

$$L_1 = L_0(x+1) = [(1,0,1,0),(1,0,0,1),(1,0,1,1),(2,0,1,0),(1,1,1,0),(1,1,0,1),(2,0,0,1),(1,1,1,1),(2,0,1,1)],$$

$$L_2 = L_0(z+1) = [(1,0,2,0),(1,0,1,1),(1,0,2,0),(2,0,2,0),(1,1,2,0),(1,1,1,1),(2,0,1,1),(1,1,2,1),(2,0,2,1)],$$

Figure 7. The partition space by the triangular prismedrons E

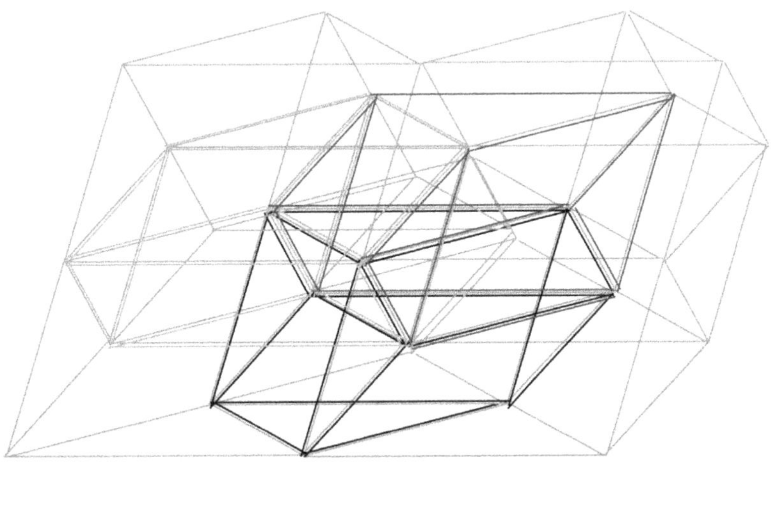

Table 3. Common elements of the triangular prismahedrons $E_0 \div E_{15}$

Edge of the Hypercube	Common Element of the Triangular Prismahedrons	Edge of the Hypercube	Common Element of the Triangular Prismahedrons
$E_0 E_1$	triangle (1,0,0,0)(1,0,1,0)(1,0,0,0)	$E_5 E_{14}$	triangle (1,1,1,1)(1,2,1,1)(2,1,1,1)
$E_0 E_2$	triangle (0,1,1,0)(0,0,1,0)(1,0,1,0)	$E_6 E_7$	triangle (1,1,1,0)(1,1,0,0)(1,1,0,1)
$E_0 E_6$	triangle (0,1,0,0)(0,1,1,0)(0,1,0,1)	$E_6 E_9$	triangle (0,1,0,1)(0,2,0,1)(1,1,0,1)
$E_0 E_8$	triangle (0,1,0,1)(0,0,0,1)(1,0,0,1)	E_7E_{10}	triangle (1,1,0,1)(2,1,0,1)(1,2,0,1)
E_1E_3	triangle (1,0,1,0)(1,1,1,0)(2,0,1,0)	$E_8 E_{11}$	triangle (1,0,0,1)(1,0,1,1)(1,0,0,2)
E_7E_1	triangle (1,1,0,0,)(1,1,1,0)(1,1,0,1)	E_8E_9	triangle (0,1,0,1)(0,1,1,1)(1,0,0,2)
E_1E_{11}	triangle (1,0,0,1)(1,1,0,1)(2,0,0,1)	E_8E_{12}	triangle (0,0,1,1)(1,0,1,1)(1,0,0,2)
E_2E_3	triangle (1,0,1,0)(1,0,2,0)(1,0,1,1)	E_9E_{10}	triangle (1,1,0,1)(1,1,0,2)(1,1,1,1)
E_2E_4	triangle (0,1,1,0)(0,1,1,1)(0,1,2,0)	E_9E_{13}	triangle (0,2,1,1)(0,1,1,1)(1,1,1,1)
E_2E_{12}	triangle (0,1,1,1)(0,0,1,1)(1,0,1,1)	$E_{11}E_{10}$	triangle (1,1,0,1)(1,1,0,2)(1,1,1,1)
E_3E_5	triangle (1,1,1,0)(1,1,2,0)(0,1,1,1)	$E_{14}E_{10}$	triangle (1,1,1,1)(1,2,1,1)(2,1,1,1)
E_3E_{15}	triangle (1,0,1,1)(1,1,1,1)(2,0,1,1)	$E_{11}E_{15}$	triangle (1,0,1,1)(1,1,1,1)(2,0,1,1)
E_4E_5	triangle (1,1,1,0)(1,1,2,0)(1,1,1,1)	$E_{13}E_{12}$	triangle (0,1,1,1)(0,1,2,1)(0,1,1,2)
E_4E_6	triangle (0,1,1,0)(0,2,1,0)(1,1,1,0)	$E_{12}E_{15}$	triangle (1,0,1,1)(1,0,2,1)(1,1,1,2)
E_4E_{13}	triangle (0,1,1,1)(1,1,1,1)(0,2,1,1)	$E_{13}E_{14}$	triangle (1,1,1,1)(1,1,2,1)(1,1,1,2)
E_7E_5	triangle (1,1,1,0)(1,2,1,0)(2,1,1,0)	$E_{14}E_{15}$	triangle (1,1,1,1)(1,1,2,1)(1,1,1,2)

Figure 8. The triangular prismahedron L

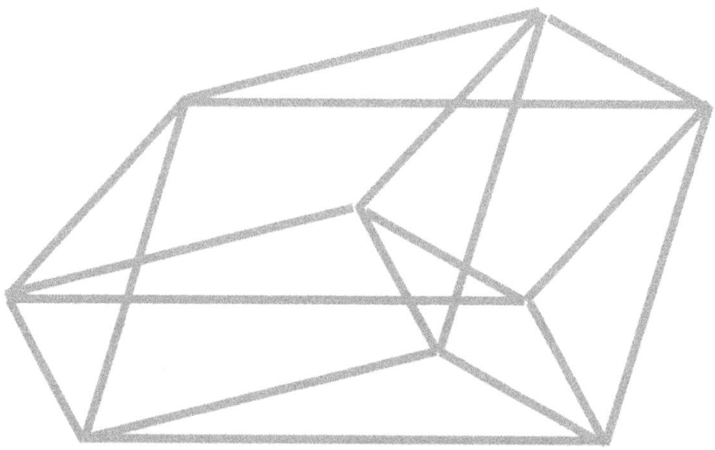

Figure 9. The triangular prismahedron C

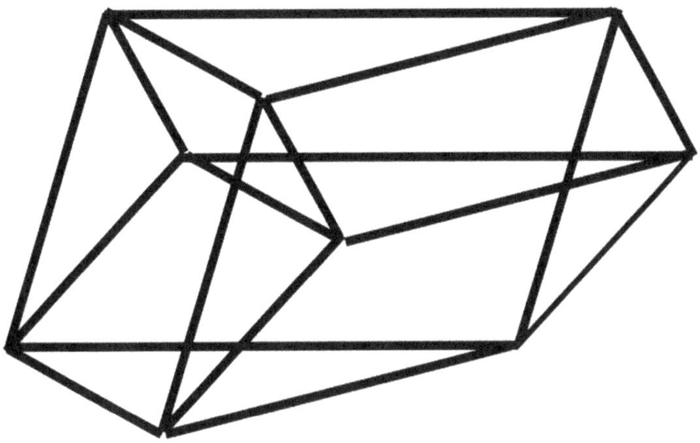

Figure 10. The triangular prismahedron D

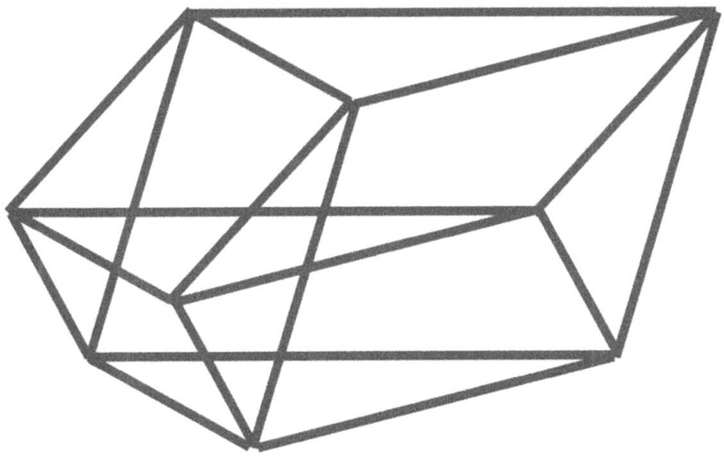

$$L_3 = L_0(x+1, z+1) = L_1(z+1) =$$
$$[(1,0,2,0),(1,0,1,1),(1,0,2,1),(2,0,2,0),(1,1,2,0,),(1,1,1,1),(2,0,1,1),(1,1,2,1),(2,0,2,1)],$$

$$L_4 = L_0(y+1, z+1) = L_2(y+1) =$$
$$[(1,1,2,0),(1,1,1,1),(1,1,2,1),(2,1,2,0),(1,2,2,0),(1,2,1,1),(2,1,1,1),(1,2,2,1),(2,1,2,1)],$$

$$L_5 = L_0(x+1, y+1, z+1) = L_3(y+1) =$$
$$[(1,1,2,0),(1,1,1,1),(1,1,2,1),(2,1,2,0),(1,2,2,0),(1,2,1,1),(2,1,1,1),(1,2,2,1),(2,1,2,1)],$$

$$L_6 = L_0(y+1) = [(0,1,1,0),(0,1,0,1),(0,1,1,1),(1,1,1,0),(0,2,1,0),(0,2,0,1),(1,1,0,1),(0,2,1,1),$$
$$(1,1,1,1)],$$

$$L_7 = L_0(y+1, x+1) = L_1(y+1) =$$
$$[(1,1,1,0),(1,1,0,1),(1,1,1,1),(2,1,1,0),(1,2,1,0),(1,2,0,1),(2,1,0,1),(1,2,1,1),(2,1,1,1)],$$

$$L_8 = L_0(t+1) =$$
$$[(0,0,1,1),(0,0,0,2),(0,0,1,2),(1,0,1,2),(0,1,1,1),(0,1,0,2),(1,0,0,2),(0,1,1,2),(1,0,1,2)],$$

$$L_9 = L_0(y+1, t+1) = L_6(t+1) =$$
$$[(0,1,1,1),(0,1,0,2),(0,1,1,2),(1,1,1,1),(0,2,1,1),(0,2,0,2),(1,1,0,2),(0,2,1,2),(1,1,1,2)],$$

$$L_{10} = L_0(x+1, y+1, t+1) = L_7(t+1) =$$
$$[(1,1,1,1),(1,1,0,2),(1,1,1,2),(2,1,1,1),(1,2,1,1),(1,2,0,2),(2,1,0,2),(1,2,1,2),(2,1,1,2)],$$

$$L_{11} = L_0(x+1, t+1) = L_1(t+1) =$$
$$[(1,0,1,1),(1,0,0,2),(1,0,1,2),(2,0,1,1),(1,1,1,1),(1,1,0,2),(2,0,0,2),(1,1,1,2),(2,0,1,2)],$$

$L_{12} = L_0(y+1, z+1) = L_2(t+1) =$
$[(1,0,2,1),(1,0,1,2),(1,0,2,2),(2,0,2,1),(1,1,2,1),(1,1,1,2),(2,0,1,2),(1,1,2,2),(2,0,2,2)],$

$L_{13} = L_0(z+1, y+1, t+1) = L_4(t+1) =$
$[(1,1,2,1),(1,1,1,2),(1,1,2,2),(2,1,2,1),(1,2,2,1),(1,2,1,2),(2,1,1,2),(1,2,2,2),(2,1,2,2)],$

$L_{14} = L_0(x+1, y+1, z+1, t+1) = L_5(t+1) =$
$[(1,1,2,1),(1,1,1,2),(1,1,2,2),(2,1,2,1),(1,2,2,1),(1,2,1,2),(2,1,1,2),(1,2,2,2),(2,1,2,2)],$

$L_{15} = L_0(x+1, t+1, z+1) = L_3(t+1) =$
$[(1,0,2,1),(1,0,1,2),(1,0,2,2),(2,0,2,1),(1,1,2,1),(1,1,1,2),(2,0,1,2),(1,1,2,2),(2,0,2,2)].$

The indicated of the triangular prismahedrons type C

$C_0 = [(0,1,0,0),(1,0,0,0),(0,1,0,1),(0,1,1,0),(1,0,1,0),(1,1,0,0),(1,0,0,1),(1,1,1,0),(1,1,0,1)],$

$C_1 = C_0(x+1) = [(1,1,0,0),(2,0,0,0),(1,1,0,1),(1,1,1,0),(2,0,1,0),(2,1,0,0),(2,0,0,1),(2,1,1,0),$
$(2,1,0,1)],$

$C_2 = C_0(z+1) = [(0,1,1,0),(1,0,1,0),(0,1,1,1),(0,1,2,0),(1,0,2,0),(1,1,1,0),(1,0,1,1),(1,1,2,0),$
$(1,1,1,1)],$

$C_3 = C_0(x+1, z+1) = C_1(z+1) =$
$[(1,1,1,0),(2,0,1,0),(1,1,1,1),(1,1,2,0),(2,0,2,0,),(2,1,1,0),(2,0,1,1),(2,1,2,0),(2,1,1,1)],$

$C_4 = C_0(y+1, z+1) = C_2(y+1) =$
$[(0,2,1,0),(1,1,1,0),(0,2,1,1),(0,2,2,0),(1,1,2,0),(1,2,1,0),(1,1,1,1),(1,2,2,0),(1,2,1,1)],$

$$C_5 = C_0(x+1, y+1, z+1) = C_3(y+1) =$$
$$[(1,2,1,0),(2,1,1,0),(1,2,1,1),(1,2,2,0),(2,1,2,0),(2,2,1,0),(2,1,1,1),(2,2,2,0),(2,2,1,1)],$$

$$C_6 = C_0(y+1) = [(2,2,0,0),(1,1,0,0),(0,2,0,1),(0,2,1,0),(1,1,1,0),(1,2,0,0),(1,1,0,1),(1,2,1,0),$$
$$(1,2,0,1)],$$

$$C_7 = C_0(y+1, x+1) = C_1(y+1) =$$
$$[(1,2,0,0),(2,1,0,0),(1,2,0,1),(1,2,1,0),(2,1,1,0),(2,2,0,0),(2,1,0,1),(2,2,1,0),(2,2,0,1)],$$

$$C_8 = C_0(t+1) =$$
$$[(0,1,0,1),(1,0,0,1),(0,1,0,2),(0,1,1,1),(1,0,1,1),(1,1,0,1),(1,0,0,2),(1,1,1,1),(1,1,0,2)],$$

$$C_9 = C_0(y+1, t+1) = C_6(t+1) =$$
$$[(0,2,0,1),(1,1,0,1),(0,2,0,2),(0,2,1,1),(1,1,1,1),(1,2,0,1),(1,1,0,2),(1,2,1,1),(1,2,0,2)],$$

$$C_{10} = C_0(x+1, y+1, t+1) = C_7(t+1) =$$
$$[(1,2,0,1),(2,1,0,1),(1,2,0,2),(1,2,1,1),(2,1,1,1),(2,2,0,1),(2,1,0,2),(2,2,1,1),(2,2,0,2)],$$

$$C_{11} = C_0(x+1, t+1) = C_1(t+1) =$$
$$[(1,1,0,1),(2,0,0,1),(1,1,0,2),(1,1,1,1),(2,0,1,1),(2,1,0,1),(2,0,0,2),(2,1,1,1),(2,1,0,2)],$$

$$C_{12} = C_0(y+1, z+1) = C_2(t+1) =$$
$$[(0,1,1,1),(1,0,1,1),(0,1,1,2),(0,1,2,1),(1,0,2,1),(1,1,1,1),(1,0,1,2),(1,1,2,1),(1,1,1,2)],$$

$$C_{13} = C_0(z+1, y+1, t+1) = C_4(t+1) =$$
$$[(0,2,1,1),(1,1,1,1),(0,2,1,2),(0,2,2,1),(1,1,2,1),(1,2,1,1),(1,1,1,2),(1,2,2,1),(1,2,1,2)],$$

$C_{14} = C_0(x+1, y+1, z+1, t+1) = C_5(t+1) =$
$[(1,2,1,1),(2,1,1,1),(1,2,1,2),(1,2,2,1),(2,1,2,1),(2,2,1,1),(2,1,1,2),(2,2,2,1),(2,2,1,2)],$

$C_{15} = C_0(x+1, t+1, z+1) = C_3(t+1) =$
$[(1,1,1,1),(2,0,1,1),(1,1,1,2),(1,1,2,1),(2,0,2,1),(2,1,1,1),(2,0,1,2),(2,1,2,1),(2,1,1,2)].$

The indicated of the triangular prismahedrons type D

$D_0 = [(0,1,1,0),(1,0,1,0),(1,0,0,1),(0,1,0,1),(0,1,1,1),(1,0,1,1),(1,1,0,1),(1,1,1,0),(1,1,1,1)],$

$D_1 = D_0(x+1) = [(1,1,1,0),(2,0,1,0),(2,0,0,1),(1,1,0,1),(1,1,1,1),(2,0,1,1),(2,1,0,1),(2,1,1,0),$
$(2,1,1,1)],$

$D_2 = D_0(z+1) = [(0,1,2,0),(1,0,2,0),(1,0,1,1),(0,1,1,1),(0,1,2,1),(1,1,1,1),(1,1,2,0),(1,1,2,1),$
$(1,0,2,1)],$

$D_3 = D_0(x+1, z+1) = D_1(z+1) =$
$[(1,1,2,0),(2,0,2,0),(2,0,1,1),(1,1,1,1),(1,1,2,1,),(2,0,2,1),(2,1,1,1),(2,1,2,0),(2,1,2,1)],$

$D_4 = D_0(y+1, z+1) = D_2(y+1) =$
$[(0,2,2,0),(1,1,2,0),(1,1,1,1),(0,2,1,1),(0,2,2,1),(1,1,2,1),(1,2,1,1),(1,2,2,0),(1,2,2,1)],$

$D_5 = D_0(x+1, y+1, z+1) = D_3(y+1) =$
$[(1,2,2,0),(2,1,2,0),(2,1,1,1),(1,2,1,1),(1,2,2,1),(2,1,2,1),(2,2,1,1),(2,2,2,0),(2,2,2,1)],$

$D_6 = D_0(y+1) = [(0,2,1,0),(1,1,1,0),(1,1,0,1),(0,2,0,1),(0,2,1,1),(1,1,1,1),(1,2,0,1),(1,2,1,0),$
$(1,2,1,1)],$

$$D_7 = D_0(y+1,x+1) = D_1(y+1) =$$
$$[(1,2,1,0),(2,1,1,0),(2,1,0,1),(1,2,0,1),(1,2,1,1),(2,1,1,1),(2,2,0,1),(2,2,1,0),(2,2,1,1)],$$

$$D_8 = D_0(t+1) =$$
$$[(0,1,1,1),(1,0,1,1),(1,0,0,2),(0,1,0,2),(0,1,1,2),(1,0,1,2),(1,1,0,2),(1,1,1,1),(1,1,1,2)],$$

$$D_9 = D_0(y+1,t+1) = D_6(t+1) =$$
$$[(0,2,1,1),(1,1,1,1),(1,1,0,2),(0,2,0,2),(0,2,1,2),(1,1,1,2),(1,2,0,2),(1,2,1,1),(1,2,1,2)],$$

$$D_{10} = D_0(x+1,y+1,t+1) = D_7(t+1) =$$
$$[(1,2,1,1),(2,1,1,1),(2,1,0,2),(1,2,0,2),(1,2,1,2),(2,1,1,2),(2,2,1,1),(2,2,1,2),(2,1,1,2)],$$

$$D_{11} = D_0(x+1,t+1) = D_1(t+1) =$$
$$[(1,1,1,1),(2,0,1,1),(2,0,0,2),(1,1,0,2),(1,1,1,2),(2,0,1,2),(2,1,0,2),(2,1,1,1),(2,1,1,2)],$$

$$D_{12} = D_0(y+1,z+1) = D_2(t+1) =$$
$$[(1,1,1,1),(2,0,1,1),(2,0,0,2),(1,1,0,2),(1,1,1,2),(2,0,1,2),(2,1,0,2),(2,1,1,1),(2,1,1,2)],$$

$$D_{13} = D_0(z+1,y+1,t+1) = D_4(t+1) =$$
$$[(0,2,2,1),(1,1,2,1),(1,1,1,2),(0,2,1,2),(0,2,2,2),(1,1,2,2),(1,2,1,2),(1,2,2,1),(1,2,2,2)],$$

$$D_{14} = D_0(x+1,y+1,z+1,t+1) = D_5(t+1) =$$
$$[(1,2,2,1),(2,1,2,1),(2,1,1,2),(1,2,1,2),(1,2,2,2),(2,1,2,2),(2,2,1,2),(2,2,2,1),(2,2,2,2)],$$

$$D_{15} = D_0(x+1,t+1,z+1) = D_3(t+1) =$$
$$[(1,1,2,1),(2,0,2,1),(2,0,1,2),(1,1,1,2),(1,1,2,2),(2,0,2,2),(2,1,1,2),(2,1,2,1),(2,1,2,2)].$$

Defining the intersections of the triangular prismahedrons in each of the sets of prismahedrons *E, L, C* or *D* from their coordinate expressions it is easy to determine that the common part of any pair of triangular prismahedrons belonging to a given set of prismahedrons (*E, L, C* or *D*) is some triangular face of the chosen pair of prismahedrons. Representing the triangular prismahedrons $L_0 \div L_{15}, C_0 \div C_{15}, D_0 \div D_{15}$ dots in three-dimensional space, for each type triangular prismahedrons can to get the hypercube. Moreover, the edges of the hypercube correspond to possible changes in the values of one of the coordinates of the vertices of the triangular prismahedrons of given type unit (see Figure 6 for the triangular prismahedrons of type *E*).

Using the coordinate expressions of the triangular prismahedrons, can establish that triangular prismahedrons belonging to different types can have common geometric elements of different dimensions that enter into triangular prismahedrons. Their intersections can be empty (that is, they do not have common elements), for example, $E_2 \cap D_1 = \emptyset$. Their intersection can be single edge, for example, $E_3 \cap L_0 = (1,0,1,0)(1,0,1,1)$,

$$E_3 \cap C_1 = (1,0,1,0)(1,1,1,0), E_2 \cap D_0 = (0,1,1,0)(1,0,1,1).$$

Their intersection can be square face, for example,

$E_7 \cap D_1 =$ (1,1,1,0)(1,1,0,1)(2,1,1,0)(2,1,0,1), $E_3 \cap D_2 =$ (1,0,2,0)(1,0,1,1)(1,1,2,0)(1,1,1,1).

Their intersection can be triangular prism also, for example,

$$E_3 \cap L_1 = (1,0,1,0)(2,0,1,0)(1,1,1,0)(1,0,1,1)(2,0,1,1)(1,1,1,1),$$

$$L_1 \cap D_0 = (1,0,1,0)(1,0,0,1)(1,0,1,1)(1,1,1,0)(1,1,0,1)(1,1,1,1).$$

Let us consider the set of the triangular prismahedrons of four types E, L, C, D of least - cubed numbers close to the origin from the considered set of the triangular prismahedrons. This set consists of the triangular prismahedrons E_0, L_0, C_0, D_0. By their coordinate expressions we establish that the pairs of the triangular prismahedrons E_0 and C_0, E_0 and L_0, C_0 and D_0, L_0 and D_0 have common trigonal prisms, and the pairs of the triangular prismahedrons E_0 and D_0, C_0 and L_0 have common planar faces of the four coal form

$$E_0 \cap C_0 = (0,1,0,0,)(1,0,0,0)(0,1,0,1)(1,0,0,1)(0,1,1,0)(1,0,1,0),$$

$$E_0 \cap L_0 = (0,0,1,0,)(0,1,1,0)(1,0,1,0)(0,0,0,1)(0,1,0,1)(1,0,0,1),$$

$$C_0 \cap D_0 = (0,1,0,1,)(1,1,0,1)(1,0,0,1)(0,1,1,0)(1,1,1,0)(1,0,1,0),$$

$$L_0 \cap D_0 = (0,1,1,1,)(1,0,1,1)(0,1,0,1)(1,0,0,1)(0,1,1,0)(1,0,1,0),$$

$$E_0 \cap D_0 = (0,1,0,1)(1,0,0,1)(0,1,1,0)(1,0,1,0),$$

$$L_0 \cap C_0 = (0,1,0,1)(1,0,0,1)(0,1,1,0)(1,0,1,0).$$

Thus, the triangular prismahedrons form closed complex of the triangular prismahedrons, which is convex polytope with dimension 4 (see Figure 11).

It is easy to see that the complex of the trigonal prismahedrons E_0, L_0, C_0, D_0 is the fundamental domain of the partition of four - dimensional space by a trigonal prismahedron, since translation of this complex with respect to the coordinates of space leads to the filling of this space without gaps. At the same time, adjacent complexes have common three - dimensional figures (rhombohedrons) included in the complex.

Figure 11. The fundamental domain of the decomposition of a four - dimensional space by a trigonal prismahedron

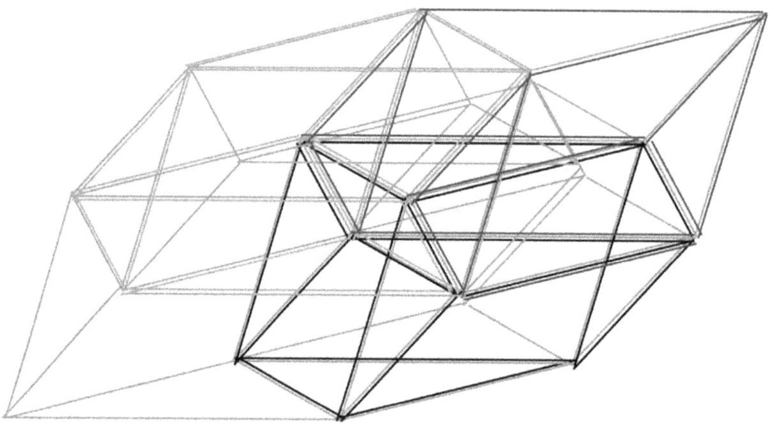

The fundamental area of the partition of a four - dimensional space by a trigonal prismahedron can be conditionally represented as a set of the trigonal prismahedrons $\frac{LD}{EC}$ in which the diagonal elements are joined to each other along two - dimensional quadrangular surfaces and the elements along the vertical and horizontally join each other connected along trigonal prism. The fundamental region with adjacent fundamental regions is bordered by rhombohedrons.

GENERAL PRINCIPLES OF THE PARTITION OF THE *N* – DIMENSIONAL SPACES BY POLYTOPIC PRISMAHEDRONS

The investigations of specific partitions of spaces by the polytopic prismahedrons allow us to formulate some general principles for partitioning spaces by the polytopic prismahedrons.

The polytopic prismahedrons, being a product of geometric figures of different dimensions, necessarily contain in their composition works of facets of figures on one - dimensional segments (edges). Therefore, they always have systems of parallel edges, and therefore parallelogons, in particular, flat faces in the form of quadrilaterals with parallel sides. In space, these parallel edges in the translation of the polytopic prismahedrons along the coordinates of space this gives rise to a system of parallel lines. Thus, between the polytopic prismahedrons in space are formed regions bounded by ribs and flat faces, i.e. again the polytopic prismahedrons are born. Their form is induced by the polytopic prismahedrons during their translation (translated polytopic prismahedrons). It either repeats the form of the translated the polytopic prismahedrons (in the case of a fairly symmetrical form of the latter) or slightly differs from their shape by the principle of "squeezing out" the shape. The number of possible deviations of free regions from translation the polytopic prismahedrons depends on the deviations from the symmetry of the most translation the polytopic prismahedrons. The contact of the polytopic prismahedrons of different shapes is carried out along prism generators (as factors), and on surfaces formed by these generators. In addition, the contact of the polytopic prismahedrons is realized by three - and multidimensional figures connected by these generators. Translated the polytopic prismahedrons form a network in space. If we introduce a coordinate system directed along the edges of some polytopic prismahedron emanating from its arbitrary vertex, taken as the origin, and the length of each edge is considered equal to one, then each vertex of the polytopic prismahedron acquires the notation. This designation consists of a sequence of zeros and ones as the values of the coordinates of the n - dimensional space. The translation of the polytopic prismahedron by the length of the edge along one of the coordinates of the space increases the value of this coordinate at all vertices of the polytopic prismahedron by one. The introduction of integer values of the coordinates of the vertices of a polytopic prismahedron allows analytically describe not only the translated polytopic prismahedron, but also polytopic prismahedrons located between the translated polytopic prismahedrons.

If we denote every translatable polytopic prismahedron by a point in n - dimensional space, then when translating this polytopic prismahedron by one in all coordinates of the n - dimensional space, then the number of these points is equal to $\sum_{i=0}^{n} !\,_{n}^{i} = 2^{n}$, where C_n^i is the number of combinations of the total number of coordinates n on the number of coordinates equal to unity . These points can be considered vertices of the n - cube, since the number of vertices in the n - cube is equal to 2^n . The number of edges in the n - cube is $n2^{n-1}$ (Zhizhin, 2014). In this case, each edge in the n - cube corresponds to translation of the polytopic prismahedron by one from one of the coordinate n – dimension space.

The analytical description of the polytopic prismahedrons allows analytically to describe the partition of an n - dimensional space by the polytopic prismahedrons. In particular, establish exactly the intersection areas of the polytopic prismahedrons during their translation. Translated the polytopic prismahedrons and polytopic prismahedrons located between translated polytopic prismahedrons naturally form the fundamental regions of partition of n - dimensional space by polytopic prismahedrons. An analytic description of the polytopic prismehedrons also allows one analytically describes the fundamental regions of partitions of n - dimensional spaces by the polytopic prismahedrons.

Due to the nature of the polytopic prismahedrons, the fundamental regions of partition of the n -dimensional spaces by the polytopic prismahedrons can contact each other along geometric elements of different dimensions: edges, flat surfaces, three - dimensional and multidimensional figures.

The presented principles of partition of n - dimensional spaces by the polytopic prismahedrons, based on the study of specific geometric constructions, radically differ from the simplified speculative representations of the theory of the stereohedrons of Delone (Delone, 1961).

THE CONNECTION BETWWEN HIERARCHICAL AND TRANSLATIONAL FILLING OF SPACE

The discovery of quasicrystals was associated with the detection of the absence in the diffraction patterns of intermetallic compounds of translational symmetry (Shechtman at al., 1984). This led to large the number of works on the study of the hierarchical and chaotic filling of the plane and space with various geometric figures, considering these ways of filling space as an opposition to the well - known method of translational space filling (Penrose, 1979; Mackay, 1981; Lord, 1991; Lord at al., 2006; Audier & Duneau, 2000; Zhizhin, 2012). However, it was later discovered that the diffraction patterns of intermetallic compounds considered as projections of structures from spaces of higher dimension have a latent periodicity (Shevchenko, Zhizhin, Mackey, 2013a, b; Zhizhin, 2014). Chapter 7 has already dealt with hierarchical filling of spaces. In this section, one show that there is a direct geometric relationship between the two types of filling the hierarchical and translational spaces. Let's return to the hierarchical filling of the two - dimensional plane with the correct pentagon, considered in Chapter 7. Following the method outlined in this chapter, continue the sides of the red pentagon to the intersection (Figure 12).

These intersections are the vertices of another regular red pentagon of a larger size. Continuing the sides of this pentagon to the intersection, can to get the vertices of an even larger pentagon. This is how the plane is filled with a regular pentagon. Now one continue the sides of the regular red pentagons into the whole foreseeable part of the plane (within the picture). One also draw part of the lines parallel to these sides. This leads to the formation of a series of intersection points that can be considered as vertices of other regular black pentagons (see Figure 12). It is found that these regular black pentagons translationally fill the plane. Thus, in Figure 12 there is also a hierarchical and translational filling of the plane by the same figures: regular pentagons.

Consider the second example.

Figure 12. Hierarchical and translational filling of the plane with a regular pentagon

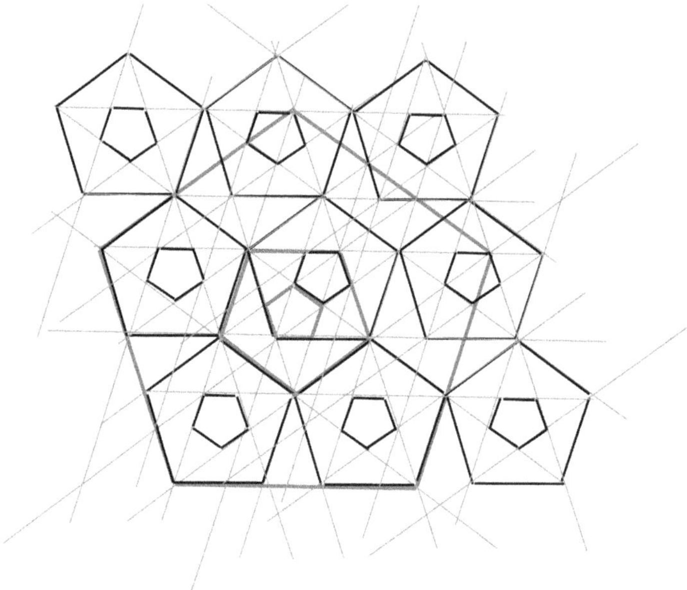

Let there be a polytopic prismahedron the 5 - simplex - prism (Figure 3 in Chapter 4). This prismahedron has dimension 5. Can to introduce the coordinate system of the five - dimensional space in one of the vertices of the 5 - simplex - prism (Figure 13 of Chapter 8).

Figure 13. Hierarchical and translational filling of the five - dimensional space by the 5 – simplex - prism

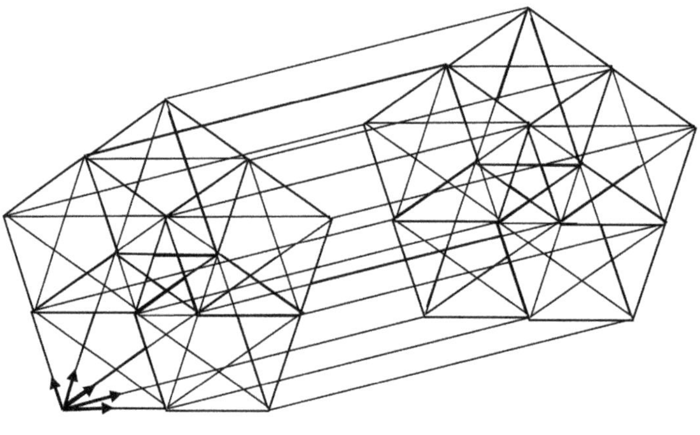

Assuming the length of all edges of the 5 - simplex - prism per unit, one will translate the 5 -simplex prism by the length of the edge along each coordinate direction within the figure. There are four the 5-simplex – prisms, which together with the original 5 - simplex - prism form a 5 -simplex - prism of a larger size. This process can be continued by combining the translation of a 5 – simplex - prism with growth its size. This is hierarchically and translationally filling the five - dimensional space with a 5 - simplex - prism on the same system of parallel lines. Thus, there is no contradiction between hierarchical and translational filling of spaces.

MODERN CONCEPT OF DESCRIBING THE STRUCTURE OF MATTER

Can to recall briefly the main concepts and results of the classical theory of discrete systems. An example of the realization of constructing a space from congruent polytopes (18 Hilbert's problem) to be the crystalline structures that are widespread in nature (Hilbert, 1901). A crystalline structure is called correct if the groups in all those space motions, like a rigid whole that combine this structure with itself, is discrete and has a fundamental domain. A group G of motions is said to be discrete, if there exists a point A and a positive number r such, that every point different from A and equivalent to A is relative to G (that is, the one into which the point A passes by the motion from G) lies no closer to A then on distance r. The fundamental domain of the group G is a set of points of space such that

1. All its points are not equivalent to each other with respect G, and
2. Any point of the space is equivalent to an equivalent point of this domain with respect G.

Such groups G are called crystallographic. These classical definitions (Delone, Padurov & Alexandrov, 1934) implicitly assume that the considered spaces are infinite, although all crystal structures in nature have a boundary and a finite volume. It was first established by Fedorov (Fedorov, 1890) and somewhat later by Schönfles (Schönfles, 1891) that there are 17 on the two-dimensional plane, and 230 crystallographic groups in the three - dimensional space. If one add to the conditions of discreteness and finiteness of the fundamental domain the requirement that in the group G there exist an n - dimensional subgroup T of parallel transfers(if G is n -dimensional), then the proof of the finiteness of the number of such groups follows directly from the Frobenius argument (Frobenius, 1911). Evidence for the existence of such n - dimensional subgroups T was obtained in a number of papers (Schönfles, 1891, Zasenkaus, 1948, Bieberbach, 1910, 1911, 1912). Delone (Delone,1937) introduced the concept about system of the distribution of discrete matter, based on two conditions:

- The existence of a finite smallest radius of the ball inside which there are no points of the system (the concept of an empty ball), and
- The existence of a radius of the ball inside which there is necessarily at least one point of the system.

The question arises as to what are the convex polytopes that play the role of fundamental domains in n - dimensional Euclidean space. Based on the introduced notion of a discrete system Delone proved (Delone, 1961) that a normal partition of a space (polytopes are adjacent to each other along entire (n - 1) - dimensional faces) for any n there are only a finite number of topologically different types of partition. In the definition of normal partitions is assumed that for every (n - 1) face of polytope P in the normal decomposition there is one and only one other polytope S having this same face. If we take the group G and repeat a point A, then we obtain a regular system of points. The Dirichlet domains of its points form some regular Dirichlet decomposition connected with the group G. It is normal. Delone and Sandakova proved (Delone, Sandakova, 1961) that the stereohedrons of these partitions for a fixed n can only be of a finite number of topologically different types. The edge grids of the stereohedrons do not stretch to each other. If one does not require the normality of partitions into convex fundamental domains, then there can be infinitely many such topologically distinct partitions (Zamorzayev, 1965).

It is clear absolutely that the concept of discrete systems introduced by Delone clearly contradicts the existence of a scaling process, which has been proved experimentally when studying phase transitions of the second kind in condensed substances. The concept of an empty ball does not allow a continuous reduction in scale. In this connection, it is necessary to harmonize the theory of describing the structure of matter with scaling processes. One consider an arbitrary n - dimensional system of points in n - dimensional space. The system is discrete, i.e. between points there is always some distance, which can only tend asymptotically to zero, not reaching an exact equality to zero. The system has in the neighborhood of each point the hierarchical distribution of points (Mackay, 2001). The latter leads to the existence of a periodic distribution of points in space.

The introduction of such a method of describing the structure of matter corresponds to the diffraction patterns of the nanostructures and to the scaling process discovered in recent years. It is essential that the lattice of points of this system allows for the separation in it, as shown in Chapter 1, of all Platonic solids, Bravais and Delone cells. Thus, this discrete system is universal. Based on this new conception of describing the structure of matter, it was possible to obtain and systematize (Chapter 4) the polytopic prismahedrons of the higher dimension, which, as shown in this Chapter, allow filling an n - dimensional space without gaps. The direct construction of the polytopic prismahedrons showed that the obligatory condition accepted in the theory of the Delone stereohedrons (Tarasov, 1997), does not hold, the separation of vertices of neighboring stereohedrons along the n -1 plane.

The discrete Delone system has so far not been able to obtain a single example of a stereohedron higher dimensional (Galiulin, 2003). Until no enumeration of the stereohedrons of Delone with dimension 3 of normal partitions has been obtained. There are only isolated examples of these stereohedrons (Shtogrin, 1973; Peter, 1981). Moreover, the use of discrete Delane systems can lead to incorrect results. For example, in the work of Ryzhkov, Shushbaev (Ryzhkov & Shushbaev, 1981) is asserted on the basis of ideas about these systems, that with the help of a 4 – cross -polytope one can obtain the correct partition of the space 4D. However, a direct construction by the methods developed in this book can show that this is not so (Zhizhin, 2017).

The most complete variants of normal monohedral, dihedral and polyhedral partitions of the two - dimensional Euclidean plane was considered by Zhizhin (Zhizhin, 1993).

The scaling process is inextricably linked with the hierarchical filling of space. In nature it corresponds, in particular, to the formation of clusters. In the hierarchical filling of space, as shown in Chapter 7, a natural restriction of the geometric analysis follows the restriction of the volume of the space occupied by such a filling. This is observed in clusters, which always have a limited volume. It is important that, as shown in Chapter 7, hierarchical filling of space can be coordinated with the translational filling of space. In this case, a model of a uniformly expanding universe appears.

Hierarchical filling of space leads to the existence of two types of clusters. In one of these types the entire volume of the cluster it is occupied by the polytope boundary complex. This corresponds to the notion that the cluster surface occupies all of its volume. In the cluster of second type boundary complex does not occupy the entire volume of the polytope, it has internal points. This corresponds to the fact that there is free space inside the cluster.

Thus, a modern description of the structure of matter leads to a new content of Hilbert's 18th problem.

CONCLUSION

One list the most significant results obtained in this book.

Areas of research into the phenomena of nature in which the influence of polytopes of higher dimension is manifested. These include studies of the structures of many chemical compounds whose molecules exhibit the properties of polytopes of higher dimension. This leads to the creation of higher - dimensional stereochemistry. Phase transitions of the second kind are accompanied by a change in the symmetry of the structure of matter, the description of which, in agreement with the experimental data, requires the attraction of spaces of higher dimension. Phase transitions of the second kind are also characterized by the scaling process, i.e. a change in the scale of the elementary cells of the corresponding structures, hence the change in the scale of polytopes of higher dimension. A system of differential equations that describes phase transitions of the second kind is qualitatively investigated. The fluctuation modes in the process of scaling are found, i.e. modes of chaotic scale of elementary cells of structures. Investigations of quasi -crystals also reveal the existence of polytopes of higher dimension, since it is in spaces of higher dimension that the diffraction patterns of quasicrystals manifest the latent translational symmetry of quasicrystals. Elementary cells of quasicrystals, having the form of the polytopic prismahedrons, are given, i.e. polytopes of higher dimension. Using polytopes of higher dimension for the first time managed to explain the geometric cause of the formation of a spiral in a DNA molecule, to obtain an analytical formula for calculating this spiral.

The structure of a polytope of dimension n consisting of points of the boundary complex including a set of faces from zero to n -1 and a set of interior points that are not belonging to the boundary complex is considered. The value is equal to the number of elements of the boundary complex, which the given element belongs, having dimension one greater than the given element of the boundary complex is denoted coefficient incidence of the given element. It is proved that the coefficient incidence of an element of dimension i of the boundary complex of an n - cube and n - simplex is equal to the difference of the dimension of the cube or simplex n and the dimension of this element i. The construction of n – cross - polytopes of various dimensions, including large dimension, revealed that the incidence coefficient of the elements of this polytope is substantially higher than the incidence coefficient of the elements of the n - cubes and the n -simplexes for equal values of the dimension of polytopes. The incidence coefficients of all three types of polytopes increase linearly with increasing n, but the proportionality coefficient of this dependence in the cross - polytopes is twice as large as the proportionality coefficient of the cubes and the simplexes. It is shown that clusters, as polytopes of higher dimension, can be of two types. In one of them, the polytope boundary complex occupies the entire volume of the polytope. This corresponds to the situation when the surface of the cluster occupies its entire volume. The second type of cluster has internal points that do not belong to the boundary complex. This corresponds to the case when the surface of the cluster does not occupy its entire volume and there are interior points below the surface.

Significantly larger values of the incidence coefficients of the elements of the boundary complex of n – cross - polytopes as compared to the n - cubes and n - simplexes should be taken into account in the study of structures involving n – cross - polytopes.

Groups of symmetry transformations of regular polytopes of dimension n (n - cubes, n - simplexes, n – cross - polytopes) are considered, using symmetry transformation groups of their facets. It is investigated how a certain symmetry transformation of the facet leads to the transformation of the symmetry of the polytope, under the condition of continuity of the polytope as a whole. It is established that the number of symmetry transformations of a regular n - polytope is equal to the product of the number of symmetry transformations of the facet of the corresponding polytope by the number of facets in this polytope.

The structure of polytopes – the polytopic prismahedrons, which are products of polytopes of lower dimensionality, is investigated. The products of polytopes do not belong to the well - studied class of simplicial polytopes, and therefore their investigations are of independent interest. Analytical dependencies characterizing the structure of the product of polytopes are obtained as a function of the structures of polytope factors. Images of a number of specific polytopic prismahedrons are obtained, tables of structures of the polytopic prismahedrons are compiled, depending on the types of polytopes of the factors. The geometric properties of the boundary complexes of the polytopic prismehedrons are investigated. The polytopic prismahedrons can be considered as a result of the chemical interaction of molecules, which, from among which there is a polytope of a certain dimension. The possibility of the multivaluedness of the incidence coefficients of geometric elements of polytopic prisms is revealed. It is shown that the polytopic prismahedra, due to their nature, as products of polytopes, ensure the filling of n - dimensional spaces during their translation, creating a structure similar to the structure of quasi - crystals.

The incidence of elements of low dimension in convex regular polytopes of dimension n with respect to elements of higher dimension up to elements of dimension n - 1 is investigated. It is shown that polytopes are dual to the polytopic prismahedrons form a new class of polytopes with simultaneously different values of the incidence of elements of low - dimensional polytopes to elements of higher dimension entering in the polytope. This new type of polytopes is called poly – incident polytopes. The existence of a previously unknown polytope consisting of one hundred tetrahedrons is established. Its election is constructed, all its constituent tetrahedrons are listed. The concept of a polytope with a factor structure whose vertices consist of polytopes of large dimension is introduced. It is shown that polytopes dual to the polytopic prismahedrons are polytopes with factor - structure, consisting from polytopes of large dimension. For the first time, analytical laws for the dependence of the incidence coefficients of the elements of the n - cubes, the n - simplexes, and the n – cross - polytopes on the dimension of the polytope are established. According to these laws, the incidence coefficients increase linearly with increasing n, especially this increase occurs in the n – cross - polytopes.

The structure of the n – cross - polytopes for large values of n with an exact enumeration of elements of various dimensions entering their boundary complexes is studied in detail. Examples of chemical compounds with the structure of the n – cross - polytopes are given. It is shown that the polytopic prismahedrons can be polytopes, which simultaneously include the n - simplexes and the n – cross - polytopes as faces. Examples and images of these polytopic prismahedrons are given. The question of a polytope of dimension 8, whose existence was announced (but not proved) by Gosset in 1900 was investigated. This polytope it was named by Coxseter a polytope of Gosset. It is established that the list of the n - simplexes and the n – cross - polytopes of different dimensions represented by Gosset as faces of this polytope contradicts the analytic laws of increasing incidence coefficients with increasing n, first established in Chapters 2, 4, 5 of this book. This testifies to the impossibility of the existence of Gosset's polytope. The examples of visualization of the Gosset polytope available in the literature are of an artistic nature and can not be proofs of its existence.

The hierarchical filling of the n - dimensional space with geometric figures is studied, accompanied by a process of discrete similar changes in their size, i.e. process of scaling. The scaling process in these fillings does not depend on time and is determined only by the geometric characteristics of the figures, which are preserved when their size is changed. Two possible ways of hierarchical filling of space are defined, under which the original figure incrementally increases its size fills the space. Investigations of the hierarchical filling of concrete geometric figures of a plane, three - dimensional space, four - and five - dimensional spaces are carried out. The denominators of geometric progressions characterizing sequences of figures in the process of scaling are determined depending on the shape of the figure and its dimension.

It is shown that the stationary Schrödinger equation describing the distribution of electrons in the vicinity of the atomic nucleus has a solution, in principle, for any dimensionality of the space around the nucleus. Previous researchers who argued that the solution of the Schrodinger equation exists only in three - dimensional space, without physical justification, changed the form of the potential energy of an electron in a space of higher dimension. As an example, a solution of the Schrödinger equation in a five - dimensional space is obtained. It is shown that the solution of the Schrödinger equation in p - dimensional space has p quantum numbers: the principal quantum number, the orbital quantum number and p - 2 magnetic quantum numbers. Taking into account the spin quantum number, the total number of quantum numbers in p - dimensional space is $p + 1$. This leads to the possibility of increasing the number of quantum cells of orbitals and, consequently, to the possibility of increasing the valence of the elements.

It is proved that polytopic prismahedrons have the necessary properties for partitioning the n -dimensional spaces of a face into a face, that is, they satisfy the conditions for solving the eighteenth Hilbert problem of the construction of n - dimensional spaces from congruent figures. General principles and an analytical method for constructing n -dimensional spaces with the help of the polytopic prismahedrons are developed. On the example of specific types of the polytopic prismahedrons (tetrahedral prism, triangular prismahedron), the possibility of such constructions is analytically proved. It was found that neighboring the polytopic prismahedrons in these constructions can have common geometric elements of any dimension less than n or do not have common elements. This refutes the speculatively accepted condition in the theory of the Delone stereohedrons that neighboring stereohedrons it is necessary to have common facets (that is, faces of dimension n -1) in or not to have common elements. It is shown that the set of lines arising in the hierarchical filling of spaces allows the existence of translational symmetry. Therefore, there are no fundamental contradictions between the hierarchical and translational filling of space, which the researchers so loved to speak of in the past. Based on the results of the research, a modern concept of describing the structure of matter is formulated, free from previously accepted conditions in the model of a discrete world. The new concept is consistent with a number of experimental data obtained in recent years on the structure of quasicrystals, phase transitions of the second kind, and so on. This leads to saturation of the eighteenth problem of Hilbert about the construction of spaces by geometric figures with a new content.

It follows from the foregoing that higher - dimensional polytopes describing chemical compounds have a number of features. This allows us to assert that they form, as a whole, a new class of polytopes of higher dimension. The geometry of these polytopes differs from the known simplicial polytopes and is of independent interest. It should be taken into account in the analysis of chemical compounds in the production of new materials, in molecular biology and nanomedicine in the treatment of diseases, molecular genetics, in analysis of plant physiology, nanotaxicology, and pharmaceuticals in the development of medicines. Obviously, the conducted investigation of the geometry of higher dimension polytopes of this class represents only the first stage in this direction.

The higher - dimensional polytopes, whose geometry has been studied in this book, are based on an analysis of the structure of substances and chemical compounds (Zhizhin, 2017). But the results obtained in the book certainly have a common scientific significance. It would be interesting to apply them, for example, to the developing area of image search and pattern recognition (Jackobczak, 2015, 2017a), in which, in particular, methods for modeling high - dimensional data are used (Jackobczak, 2016, 2017b).

REFERENCES

Audier, M., & Duneau, M. (2000). Icosahedral quasiperiodic pacing of fibers parallel to fivefold and threefold axes. *Acta Crystallographica. Section A, Foundations of Crystallography*, *56*(1), 49–51. doi:10.1107/S0108767399010533 PMID:10874416

Bieberbach, L. (1910). Uber die Bewegungsruppen des *n* - dimensionalen Euklidischen Räumes mit einem endlichen fundamentalbereich. *Gött. Nachr.*, 75 – 84.

Bieberbach, L. (1911). Uber die Bewegungsruppen der Euklidischen Räume. I. *Mathematische Annalen*, *70*(3), 207–336. doi:10.1007/BF01564500

Bieberbach, L. (1912). Uber die Bewegungsruppen des Euklidischen Räume. II. *Mathematische Annalen*, *72*, 400–412. doi:10.1007/BF01456724

Delone, B. (1937). Geometry of positive quadratic forms. *Uspekhi Matematicheskhi Nauk*, *3*, 16–62.

Delone, B. (1961). Proof of the main theorem of the theory of stereohedrons. *Reports of the Academy of Sciences of the USSR*, *138*(6), 1270–1272.

Delone, B. (1969). To the eighteenth problem of Hilbert. In P. Aleksandrov (Ed.), *Problems of Hilbert* (pp. 200–203). Moscow: Science.

Delone, B., Padurov, N., & Aleksandrov, A. (1934). *Mathematical foundations of the structural analysis of crystals*. Leningrad, Moscow: ONTI.

Delone, B., & Sandakova, N. (1961). Theory of stereohedrons. *Proceedings of the Mathematical Institute. V.A. Steklov, 64,* 28-51.

Eiji, A., Yanfa, Y., & Pennycook, S. J. (2004). Quasicrystals as cluster aggregates. *Nature Materials*, *3*(11), 759–767. doi:10.1038/nmat1244 PMID:15516956

Fedorov, E. S. (1885). *The beginning of the doctrine of figures*. Saint - Petersburg: Academy of Sciences.

Fischer, M. E., & Pfeuty, P. (1972). Critical behavior of the anisotropic n – vector model. *Physical Review B: Condensed Matter and Materials Physics*, *6*(5), 1889–1891. doi:10.1103/PhysRevB.6.1889

Frobenius, G. (1911). Ueber die unzerlegbaren diskreten Bewegungsgruppen. *Sitzb. Preuss. Akad. Wiss., Phys.- Math. K*, *1*, 654–655.

Galiulin, R. (2003). Systems B.N. Delone as the mathematical foundation of a discrete world. *J. Computed. Mat. and Math. Physics*, *43*(6), 790–801.

Hilbert, D. (1901). *Gesammelte Abhandlungen*.

Jakobczak, D. J. (2015). Application of Curve Interpolation in Data Modeling and Restoration. *IJCSIA*, *5*(2), 121–124.

Jakobczak, D. J. (2016). Modeling of High- Dimensional Data for Applications of Image Segmentation. Switzerland: Springer.

Jakobczak, D. J. (2017a). *Probabilistic Nodes Combination (PNC) for Object Modeling and Contour Reconstruction*. Hershey, PA, USA: IGI Global. doi:10.4018/978-1-5225-2531-8

Jakobczak, D.J. (2017b). Modeling of High – dimensional Data in Object Recognition. *IJIREC*, 4(1).

Kadanoff, L. P. (1966). Scaling Laws For Ising Models Near τ_c^*. *Physics*, *2*(6), 263–272. doi:10.1103/PhysicsPhysiqueFizika.2.263

Krapivsky, P. L., Redner, S., & Ben-Naim, E. (2010). *A Kinetic View of Statistical Physics*. Cambridge: Cambridge University Press. doi:10.1017/CBO9780511780516

Lord, E. A. (1991). Quasicrystals and Penrose patterns. *Current Science*, *61*, 313–319.

Lord, E. A. (2006). *New Geometry for New Materials*. Cambridge: Cambridge University Press.

Mackay, A. (2001). On Complexity. *Crystallography Reports*, *46*(4), 524–526. doi:10.1134/1.1387117

Mackay, A. (2001). On Complexity. *Crystallography Reports*, *46*(4), 524–526. doi:10.1134/1.1387117

Mukhopadhyay, N.K., Weatherly, G.C., Lloyd, D.J. & Embury, J.D. (1993). Diffraction studies of icosahedral phases in $Al_{70}Fe_{20}W_{10}$. *Journal of Non-Crystalline Solids*, *153-154*, 1193-1197.

Penrose, R. (1974). The role of aesthetics in pure and applied mathematical research. *Bulletin - Institute of Mathematics and Its Applications*, *10*, 266–271.

Peter, E. (1986). *Geometric Crystallography*. Dordrecht: Springer.

Reinhardt, K. (1928). Zur Zerlegung der Euklidischen Raume in Kongruente Polytope. *Sitzb. Preuss. Akad. Wiss*. (pp. 150-155).

Ryzhkov, S. S., & Shusbaev, S. Sh. (1981). The structur of the L-partition for the second perfect lattice. *Mat. Collection*, *146*(2), 218–231.

Schönfless. (1891). *Kristallsysteme und Kristallstruktur*. Liepzig.

Shechtman, D., Blech, I., Gratias, D., & Cahn, J. W. (1984). Metallic phase with longerange orientational order and no translational symmetry. *Physical Review Letters, 53*(20), 1951–1953. doi:10.1103/PhysRevLett.53.1951

Shevchenko, V., Zhizhin, G., & Mackay, A. (2013). On the structure of the quasicrystals in the high dimension space. In M. V. Diudea (Ed.), *Diamond and Related Nanostructures*. Dordrecht: Springer. doi:10.1007/978-94-007-6371-5_17

Shevchenko, V.Ya., Zhizhin, G.V. & Mackay, A.L. (2013b). On the structure of quasicrystals in a space of higher dimension, *Izvestiya RAS. Chemical series, 2*, 269 - 274.

Stogrin, M. I. (1973). Correct Dirichlet-Voronoi decompositions for the second triclinic group. *Trudy Mat. Institute of Steklov, 123*, 128.

Tarasov, A.S. (1997). Complexity of convex stereohedrons. *Mathematic notes, 61*(5), 797-800.

Wilson, R. G. (1971). Renormmalization group and critical phenomena. I. Renormalization group and the Kadanoff scaling picture. *Physical Review B: Condensed Matter and Materials Physics, 4*, 3174–3183. doi:10.1103/PhysRevB.4.3174

Zamorzayev, A.M. On abnormal regular partitions of Euclidean space. *Reports of the Academy of Sciences of the USSR, 161*(1), 30-31.

Zang, X., & Kelton, K. (1993). High-order crystal approximant alloys $Ti_{54}Zr_{26}Ni_{20}$. *Journal of Non-Crystalline Solids, 153-154*, 114–118. doi:10.1016/0022-3093(93)90325-R

Zhizhin, G. V. (1993). *Parquetages of equal triangles, adjoining on whole sides*. St. Petersburg: Polytechnic.

Zhizhin, G. V. (2010). *Geometrical bases of the dissipative structures*. St. Petersburg: Polytechnic.

Zhizhin, G. V. (2012, October). Hierarchical filling of spaces with polytopes. *Paper presented at 7th St.- Petersburg meeting of Nobel Prize laureates*, St. Petersburg, Russia.

Zhizhin, G. V. (2014a). The fractal nature of disproportionate phases. *Materials (Basel)*, 10.

Zhizhin, G. V. (2014b). Incommensurable and fluctuating structures in the terrestrial space. *Biosphere*, *3*, 211–221.

Zhizhin, G. V. (2014c). *World – 4D*. St. – Petersburg: Polytechnic Service.

Zhizhin, G. V. (2014d). On the higher dimension in nature. *Biosphere*, *6*(4), 313–318.

Zhizhin, G. V. (2015, November). Polytopic prismahedrons – fundamental regions of the n-dimension nanostructures. *Paper presented at The International conference "Nanoscience in Chemistry, Physics, Biology and Mathematics,"* Cluj – Napoca, Romania.

Zhizhin, G. V. (2017). *Chemical Compound Structures and Higher Dimension of Molecules: Emerging Research and Opportunities*. Hershey, PA: IGI Global.

Zhizhin, G. V., & Diudea, M. V. (2016). Space of Nanoworld. In M. V. Putz & M. C. Mirica (Eds.), *Sustainable Nanosystems, Development, Properties, and Applications* (pp. 214–236). Hershey, PA: IGI Global.

Zhizhin, G. V., Khalaj, Z., & Diudea, M. V. (2016). Geometrical and topology dimensions of the diamond. In A. R. Ashrafi & M. V. Diudea (Eds.), *Distance, symmetry and topology in carbon nanomaterials* (pp. 167–188). London: Springer. doi:10.1007/978-3-319-31584-3_12

KEY TERMS AND DEFINITIONS

New Paradigm of Discrete *n* – Dimension World: The elementary cells of the translational filling of the *n*-dimensional space are the polytopic prismahedrons - the stereohedrons, of which Delone spoke, but he did not give a single concrete example of stereohedron. Polytopic prismahedrons, filling the *n*-dimensional space, as shown by direct construction, can have common elements in the entire range of dimensions up to the dimension of the facets, or do not have any common elements. In addition to the translational filling of the *n*-dimensional space, there is a hierarchical filling of the space, which is inextricably linked with the scaling process, i.e. discrete scale change of the figure. Translational filling of space can be combined with hierarchical filling of space. In this case, in principle, in each point of space there is an asymptotic decrease or increase in the scale of the figure (as an expansion of the Universe from each of its points). Delone's provisions on an "empty" ball, the finite minimum and maximum distances between the points of a discrete system, are not used. Thus, Hilbert's problem acquires a completely new content.

The 5 – Simplex - Prism: The product of a 4 - simplex by a segment.

The Congruent Polyhedrons: Polyhedrons that are compatible with motion.

The Polytopic Prismahedron: The product of polytopes.

The Tetrahedral Prism: The product of a tetrahedron by a segment.

The Triangular Prismahedron: The product of a triangle by a triangle.

Conclusion

This conclusion lists the most significant results obtained from this book.

Areas of research into the phenomena of nature in which the influence of polytopes of higher dimension is manifested. These include studies of the structures of many chemical compounds whose molecules exhibit the properties of polytopes of higher dimension. This leads to the creation of higher - dimensional stereochemistry. Phase transitions of the second kind are accompanied by a change in the symmetry of the structure of matter, the description of which, in agreement with the experimental data, requires the attraction of spaces of higher dimension. Phase transitions of the second kind are also characterized by the scaling process, i.e. a change in the scale of the elementary cells of the corresponding structures, hence the change in the scale of polytopes of higher dimension. A system of differential equations that describes phase transitions of the second kind is qualitatively investigated. The fluctuation modes in the process of scaling are found, i.e. modes of chaotic scale of elementary cells of structures. Investigations of quasi -crystals also reveal the existence of polytopes of higher dimension, since it is in spaces of higher dimension that the diffraction patterns of quasicrystals manifest the latent translational symmetry of quasicrystals. Elementary cells of quasicrystals, having the form of polytopic prismahedrons, are given, i.e. polytopes of higher dimension.

The structure of a polytope of dimension n consisting of points of the boundary complex including a set of faces from zero to n -1 and a set of interior points that are not belonging to the boundary complex is considered. The value is equal to the number of elements of the boundary complex, which the given element belongs, having dimension one greater than the given element of the boundary complex is denoted coefficient incidence of the given element. It is proved that the coefficient incidence of an element of dimension i of the boundary complex of an n - cube and n - simplex is equal to the difference of the dimension of the cube or simplex n and the dimension of this element i. The construction of n – cross - polytopes of various dimensions, including large dimension, revealed that the incidence coefficient of the elements of this polytope is substantially higher than the incidence coefficient of the elements of n - cubes and n -simplexes for equal values of the dimension of polytopes. The incidence coefficients of all three types of polytopes increase linearly with increasing n, but the proportionality coefficient of this dependence in cross polytopes is twice as large as the proportionality coefficient of cubes and simplexes.

Significantly larger values of the incidence coefficients of the elements of the boundary complex of n – cross - polytopes as compared to the n - cubes and n - simplexes should be taken into account in the study of structures involving n – cross - polytopes.

Groups of symmetry transformations of regular polytopes of dimension n (n - cubes, n - simplexes, n – cross - polytopes) are considered, using symmetry transformation groups of their facets. It is investigated how a certain symmetry transformation of the facet leads to the transformation of the symmetry of the polytope, under the condition of continuity of the polytope as a whole. It is established that the number of symmetry transformations of a regular n - polytope is equal to the product of the number of symmetry transformations of the facet of the corresponding polytope by the number of facets in this polytope.

The structure of polytopes - polytopic prismahedrons, which are products of polytopes of lower dimensionality, is investigated. The products of polytopes do not belong to the well - studied class of simplicial polytopes, and therefore their investigations are of independent interest. Analytical dependencies characterizing the structure of the product of polytopes are obtained as a function of the structures of polytope factors. Images of a number of specific polytopic prismahedrons are obtained, tables of structures of polytopic prismahedrons are compiled, depending on the types of polytopes of the factors. The geometric properties of the boundary complexes of polytopic prismehedrons are investigated. Polytopic prismahedrons can be considered as a result of the chemical interaction of molecules, which, from among which there is a polytope of a certain dimension. The possibility of the multivaluedness of the incidence coefficients of geometric elements of polytopic prisms is revealed. It is shown that polytopic prismahedra, due to their nature, as products of polytopes, ensure the filling of n - dimensional spaces during their translation, creating a structure similar to the structure of quasi - crystals.

The incidence of elements of low dimension in convex regular polytopes of dimension n with respect to elements of higher dimension up to elements of dimension n - 1 is investigated. It is shown that polytopes are dual to polytopic prismahedrons form a new class of polytopes with simultaneously different values of the incidence of elements of low - dimensional polytopes to elements of higher dimension entering in the polytope. This new type of polytopes is called poly – incident polytopes. The existence of a previously unknown polytope consisting of one hundred tetrahedrons is established. Its election is constructed, all its constituent tetrahedrons are listed. The concept of a polytope with a factor structure whose vertices consist of polytopes of large dimension is introduced. It is shown that polytopes dual to polytopic prismahedrons are polytopes with factor - structure, consisting from polytopes of large dimension. For the first time, analytical laws for the dependence of the incidence coefficients of the elements of the n - cubes, n - simplexes, and n – cross - polytopes on the dimension of the polytope are established. According to these laws, the incidence coefficients increase linearly with increasing n, especially this increase occurs in n – cross - polytopes.

The structure of n – cross - polytopes for large values of n with an exact enumeration of elements of various dimensions entering their boundary complexes is studied in detail. Examples of chemical compounds with the structure of n – cross - polytopes are given. It is shown that polytopic prismahedrons can be polytopes, which simultaneously include n - simplexes and n – cross - polytopes as faces. Examples and images of these polytopic prismahedrons are given. The question of a polytope of dimension 8, whose existence was announced (but not proved) by Gosset in 1900 was investigated. This polytope was named by Coxseter a polytope of Gosset. It is established that the list of n - simplexes and n – cross - polytopes of different dimensions represented by Gosset as faces of this polytope contradicts the analytic laws of increasing incidence coefficients with increasing n, first established in Chapters 2, 4, 5 of this paper. This testifies to the impossibility of the existence of Gosset's polytope. The examples of visualization of the Gosset polytope available in the literature are of an artistic nature and cannot be proof of its existence.

The hierarchical filling of the n - dimensional space with geometric figures is studied, accompanied by a process of discrete similar changes in their dimensions, i.e. process of scaling. The scaling process in these fillings does not depend on time and is determined only by the geometric characteristics of the figures, which are preserved when their size is changed. Two possible ways of hierarchical filling of space are defined, under which the original figure incrementally increases its size fills the space. Investigations of the hierarchical filling of concrete geometric figures of a plane, three-dimensional space, four - and five - dimensional spaces are carried out. The coefficients of geometric progressions characterizing sequences of figures in the process of scaling are determined depending on the shape of the figure and its dimension.

It is proved that polytopic prismahedrons have the necessary properties for partitioning the n -dimensional spaces of a face into a face, that is, they satisfy the conditions for solving the eighteenth Hilbert problem of the construction of n - dimensional spaces from congruent figures. General principles and an analytical method for constructing n -dimensional spaces with the help of polytopic prismahedrons are developed. On the example of specific types of polytopic prismahedrons (tetrahedral prism, triangular prismahedron), the possibility of such constructions is analytically proved. It was found that neighboring polytopic prismahedrons in these constructions can have common geometric elements of any dimension less than n or do not have common elements. This refutes the speculatively accepted condition in the theory of the Delone stereohedrons that neighboring stereohedrons it is necessary to have common facets (that is, faces of dimension n -1) in or not to have common elements. It is shown that the set of lines arising in the hierarchical filling of spaces allows the existence of translational symmetry. Therefore, there are no fundamental contradictions between the hierarchical and translational filling of space, which the researchers so loved to speak of in the past.

It is shown that the stationary Schrödinger equation describing the distribution of electrons in the vicinity of the atomic nucleus has a solution, in principle, for any dimensionality of the space around the nucleus. Previous researchers who argued that the solution of the Schrodinger equation exists only in three - dimensional space, without physical justification, changed the form of the potential energy of an electron in a space of higher dimension. As an example, a solution of the Schrödinger equation in a five - dimensional space is obtained. It is shown that the solution of the Schrödinger equation in p - dimensional space has p quantum numbers: the principal quantum number, the orbital quantum number and p - 2 magnetic quantum numbers. Taking into account the spin quantum number, the total number of quantum numbers in p - dimensional space is $p + 1$. This leads to the possibility of increasing the number of quantum cells of orbitals and, consequently, to the possibility of increasing the valence of the elements.

It follows from the foregoing that higher-dimensional polytopes describing chemical compounds have a number of features. This allows us to assert that they form, as a whole, a new class of polytopes of higher dimension. The geometry of these polytopes differs from the known simplicial polytopes and is of independent interest. It should be taken into account in the analysis of chemical compounds in the production of new materials, in molecular biology and nanomedicine in the treatment of diseases, molecular genetics, in analysis of plant physiology, nanotaxicology, and pharmaceuticals in the development of medicines. Obviously, the conducted investigation of the geometry of higher dimension polytopes of this class represents only the first stage in this direction.

Index

M

N

O

P

Q

S

T

Z

www.ingramcontent.com/pod-product-compliance
Lightning Source LLC
LaVergne TN
LVHW081248011026
849162LV00004BA/853